THE

AMERICAN ALLIGATOR

http://ulpress.org
University of Louisiana at Lafayette Press
P.O. Box 43558
Lafayette, LA 70504-3558

Library of Congress Cataloging-in-Publication Data

Names: Hastings, Robert W. author
Title: The American alligator : abused, protected, restored / Robert W.
    Hastings.
Description: Lafayette, LA : University of Louisiana at Lafayette Press,
    2025. | Includes bibliographical references.
Identifiers: LCCN 2025008547 | ISBN 9781959569237 paperback
Subjects: LCSH: American alligator--Conservation--United States | American
    alligator--Conservation--History | Endangered species--Breeding--United
    States
Classification: LCC QL666.C924 H37 2025
LC record available at https://lccn.loc.gov/2025008547

cover image: *Basking American Alligator* by Juan Gracia, courtesy of Shutterstock

THE
AMERICAN ALLIGATOR

Abused, Protected, Restored

Robert W. Hastings

2025
University of Louisiana at Lafayette Press

Table of Contents

List of Tables and Figures

Acknowledgments

This book is dedicated to the outstanding alligator research team of Ted Joanen and Larry McNease of the Louisiana Wildlife and Fisheries Commission, whose impressive collection of research reports provided the basis for much of this book. I also thank them for providing me with reprints of most of their publications. I thank Dr. Ruth B. Elsey, also of the Louisiana Wildlife and Fisheries Commission, who has continued alligator research at the Rockefeller Wildlife Refuge at Grand Chenier, Louisiana, and who reviewed and offered excellent suggestions to improve an early version of this manuscript. I thank Steve Platt and Chris Brantley, who, as graduate students at Turtle Cove Environmental Research Station, stimulated my initial interest in alligator research. I also acknowledge the extensive knowledge, advice, and assistance of Mr. Hayden Reno, station manager at Turtle Cove, whose years of experience in the swamps, marshes, and waters of the upper Lake Pontchartrain Estuary proved invaluable in guiding much of the activity at the research station. I thank Dr. Shane K. Bernard, historian at Avery Island, Louisiana, who reviewed an early manuscript of this book and made many excellent recommendations to improve it and also provided the historical photographs of Mr. Edward McIlhenny. I also thank Ms. Devon Lord (editor in chief) and the University of Louisiana at Lafayette Press for their work in preparing this book for publication.

Chapter One

Introduction

The American alligator is one of the most iconic animal species in North America. It is the keystone species in freshwater swamps and marshes of the Southeast, and as such, has significant ecological value in these habitats. Alligators have provided an important source of food for many Indigenous peoples and have also represented important spiritual symbols and totemic figures for many Native cultures in North America. With the growth of European immigrant populations in the 1800s, alligators became increasingly valued for their skins. As often happens with natural resources, greed generated excessive exploitation and alligator populations became severely depleted by the late 1800s and early 1900s. This abuse continued through the middle of the twentieth century, when concerned wildlife biologists and environmentalists came to the rescue and ended the alligators' population decline. This book is intended to trace the history and success of that conservation effort to ensure the survival of this valued, living dinosaur-like reptile.

At one time, the American alligator was abundant in freshwater environments of southeastern North America's coastal plain, and paleontologists have dated fossils of extinct crocodilian species back to the Cretaceous Period some 100 million years ago. However,

American alligator at Payne's Prairie, Florida, February 18, 2010[1]

1.   Unless otherwise noted, all photographs taken by the author.

accurately estimating prehistoric population numbers and range of alligators is impossible and likely varied significantly with natural factors such as climate variations and sea level rise that affected the amount of coastal wetlands habitat. Beginning in the 1800s, human factors such as habitat destruction and excessive hunting for hides reduced alligator numbers dramatically—reaching threateningly low levels by the early 1900s. If this trend had continued, extinction could have been a possibility, if not a probability. Such a possible fate is amazing, as this large, formidable reptile had few, if any, natural enemies as adults. Moreover, the alligator occupies aquatic and semi-aquatic freshwater habitats somewhat isolated from the upland terrestrial habitats occupied by humans—who became its greatest enemy. Yet the coming of the "modern" world, with its exploding human population of European immigrants, meant that the natural habitat available to support populations of alligators and other wild animals decreased rapidly. In addition, many native animal species, including alligators, were excessively killed for human use.

The 1800s began a period of tragic environmental damage in North America. As described by William T. Hornaday in his 1913 conservation classic *Our Vanishing Wildlife*, "No wild species of bird, mammal, reptile or fish can withstand exploitation for commercial purposes."[2] Hornaday further emphasized that this principle, which he considered "an inexorable law of Nature," even applies to species in remote habitats that might provide some protection from pursuit by human hunters: "Even the whales of the deep sea, the walrus of the arctic regions, the condors of the Andes and alligators of the Everglade morasses are no exception to the universal rule."[3]

During this period, economic value plus greed without limits stimulated uncontrolled and rampant slaughter of such species as the passenger pigeon, snowy egret, and, perhaps most visibly, the American bison. The passenger pigeon was once considered the most abundant bird in North America but became extinct in 1914 as a result of excessive shooting for commercial food markets. The snowy egret was hunted almost to extinction in the 1800s for the sole purpose of obtaining its feathers, desired to decorate women's hats. It was saved in no small part by the early conservation efforts of Mr. Edward A. McIlhenny (see Chapter 6), noted alligator observer of Avery Island, Louisiana, and son of Tabasco creator Edmund McIlhenny.

---

2. William T. Hornaday, *Our Vanishing Wildlife: Its Extermination and Preser*vation (C. Scribner's Sons, 1913), 63.

3. Hornaday, *Our Vanishing Wildlife*, 63. We should qualify this rule to mean "uncontrolled" exploitation for commercial purposes.

Estimates suggest that the American bison numbered some thirty million prior to the 1800s, but fewer than 1,000 were thought to survive the extreme slaughter by the end of the century. Dedicated conservation efforts saved the species from extinction, though it will never become as abundant as it once was because most of its prairie habitat no longer exists.

The American alligator suffered similar abuse through the early 1900s before it was eventually recognized as potentially endangered and successful efforts were initiated to protect and enhance its populations and habitats. Otherwise, it could very well have experienced a similar fate as the passenger pigeon. As Edward McIlhenny wrote, "That the alligator has already been exterminated over a large portion of its former habitat is a fact, and one that civilization should not be proud of."[4] Over his life, Mr. McIlhenny became a significant authority on alligators and a leading champion for their conservation and habitat protection in Louisiana.[5]

The justifications for protecting alligators to maintain and sustain a viable population for future generations are many. They have long been valued for providing people with food and stimulating their palates with alligator recipes, enhancing fashions with alligator leather products, and for being a main attraction in zoos and other animal tourist attractions. A chapter in a 1931 bulletin from the Louisiana Department of Conservation noted that alligators had historically been an important article of commerce in the state and provided an occupation and livelihood for thousands of people in coastal lowlands.[6]

However, their true value is much more significant. Alligators are a keystone species in rivers, marshes, and swamps in that they greatly affect the habitat for other species that may depend upon them for survival. They create "gator holes" and dens, which provide aquatic habitat for other animals, especially important during drought when shallower parts of the wetland become completely dry.[7] Their mounds of vegetation created for nesting and the elevated spoil banks that may form adjacent to their dens provide slightly

4. E. A. McIlhenny, *The Alligator's Life History* (Christopher Publishing House, 1935), 116.
5. In fact, McIlhenny's book was described by Dr. Archie Carr (in the foreword to the republished edition) as "the best existing account of the natural history of one of the world's most versatile reptiles." See Archie Carr, "Foreword," in *The Alligator's Life History* by E. A. McIlhenny (Society for the Study of Amphibians and Reptiles, 1976).
6. Stanley C. Arthur, "The Alligator," in *Bulletin 18 (Revised): The Fur Animals of Louisiana*, 165–86 (Louisiana Department of Conservation, 1931).
7. Archie Carr, "Alligators: Dragons in Distress," *National Geographic* 131, no. 1 (1967): 133–48; L. Brandt, M. Campbell, and F. Mazzotti, "Spatial Distribution of Alligator Holes in the Central Everglades," *Southeastern Naturalist* 9, no. 3 (2010): 487–96.

Alligator nest of mounded marsh vegetation at
Manchac Wildlife Management Area, Louisiana, July 7, 1987

higher ground in low marshy environments, where other species may rest or nest. Alligators are the top predator in many wetlands, thus controlling the populations of their prey species. There is also a significant aesthetic value in alligators, resulting in their being a main attraction for naturalists, environmentalists, and swamp tourists.

Chapter Two

Classification of Alligators and Other Crocodilians

Populations of American alligators have occupied rivers and wetlands of the southeast portion of North America for millions of years. Even today, alligators and their close relatives, the crocodiles, gavials, and caimans (Table 1), are the largest living reptiles. As such, they are an impressive and significant part of our environment.

Biologists today classify the American alligator as follows:
Class: Reptilia
Subclass: Archosauria
Order: Crocodylia
Family: Alligatoridae
Genus and Species: *Alligator mississippiensis*

Its closest living relative is the Chinese alligator (*Alligator sinensis*), found only on the other side of the world, but critically endangered. Also included in the family Alligatoridae are six species of caimans native to Central and South America, one of which, the spectacled caiman (*Caiman crocodilus*), has established a breeding population in south Florida, apparently originating from released pets.

Closely related to the alligators are crocodiles and gavials, which are also included in the order Crocodylia. (All members of this order are also known as crocodilians.) Classified in the family Crocodylidae are sixteen species of crocodiles distributed worldwide in the tropics, including the American crocodile (*Crocodylus acutus*), also a resident of south Florida. The two slender-snouted forms known as gavials or gharials, classified in the family Gavialidae, occur in Southeast Asia and are also considered critically endangered.

Table 1. Species of Living Crocodilians

Key: CD: Conservation Dependent; CR: Critically Endangered; LC: Least Concern; LR: Least Risk; VU: Vulnerable

FAMILY AND COMMON NAME	SPECIES	NATIVE RANGE	IUCN CATEGORY
ALLIGATORIDAE			
American Alligator	*Alligator mississippiensis*	SE North America	LR/LC
Chinese Alligator	*Alligator sinensis*	East China	CR
Spectacled Caiman	*Caiman crocodilus*	Central & South America	LR/LC
Broad-Snouted Caiman	*Caiman latirostris*	Southeast South America	LR/LC
Yacare Caiman	*Caiman yacare*	Central South America	LR/LC
Black Caiman	*Melanosuchus niger*	Amazon Basin of South America	LR/CD
Cuvier's Dwarf Caiman	*Palaeosuchus palpebrosus*	North South America	LR/LC
Schneider's Smooth-Fronted Caiman	*Palaeosuchus trigonatus*	North South America	LR/LC
CROCODYLIDAE			
American Crocodile	*Crocodylus acutus*	Central & NW South America, West Indies, & Extreme South Florida	VU
West African Slender-Snouted Crocodile	*Mecistops cataphractus*	West Africa	CR
Central African Slender-Snouted Crocodile[1]	*Mecistops leptorhynchus*	Central Africa	Not assessed

1. Additional species recognized by Kent Vliet, *Alligators: The Illustrated Guide to Their Biology, Behavior, and Conservation* (Johns Hopkins Press, 2020).

Orinoco Crocodile	*Crocodylus intermedius*	Orinoco River of South America	CR
Australian Freshwater Crocodile	*Crocodylus johnstoni*	North Australia	LR/LC
Philippine Crocodile	*Crocodylus mindorensis*	Philippines	CR
Morelet's Crocodile	*Crocodylus moreletii*	Central America	LR/LC
Nile Crocodile	*Crocodylus niloticus*	Africa	LR/LC
West African Crocodile	*Crocodylus suchus*	West & Central Africa	Not assessed
New Guinea Freshwater Crocodile	*Crocodylus novaeguineae*	New Guinea & Indonesia	LR/LC
Mugger Crocodile	*Crocodylus palustris*	South Central Asia	VU
Saltwater Crocodile	*Crocodylus porosus*	South Central Asia, East Indies, & Australia	LR/LC
Cuban Crocodile	*Crocodylus rhombifer*	Cuba	CR
Siamese Crocodile	*Crocodylus siamensis*	Indochina Peninsula & Indonesia	CR
African Dwarf Crocodile	*Osteolaemus tetraspis*	West & Central Africa	VU
Congo Dwarf Crocodile[2]	*Osteolaemus osborni*	Central Africa	Not assessed
GAVIALIDAE			
Indian Gharial	*Gavialis gangeticus*	India & Nepal	CR
Sunda Gharial (False Gharial)	*Tomistoma schlegelii*	Malaysia & Indonesia	VU

Source: Crocodile Specialist Group, http://www.iucncsg.org/pages/Conservation-Status.html.

2. Additional species recognized by Kent Vliet, *Alligators: The Illustrated Guide.*

American crocodile at (*above*) Philadelphia Zoo, March 10, 1975 and (*below*) Ross Allen Reptile Institute, Silver Springs, Florida, July 25, 1970

Indian Gharial at St. Augustine Alligator Farm Zoological Park, April 8, 2021

Five species of non-native crocodilians, which could potentially compete with indigenous species, have been introduced to Florida since 1960, but only the spectacled caiman has thus far established a breeding population.[3] The other four are the slender-snouted crocodile (*Mecistops cataphractus*), Cuvier's dwarf caiman (*Paleosuchus palpebrosus*), Schneider's smooth-fronted caiman (*Paleosuchus trigonatus*), and Nile crocodile (*Crocodylus niloticus*). The Nile crocodile is of special concern because of its large size, posing potential danger to people and other animals, and also because it could threaten native crocodilians through predation, competition, and hybridization. Research has suggested that the Nile crocodile could survive throughout coastal Florida and along the whole coastline of the Gulf of Mexico. This area includes the entire Florida range of the American crocodile, which could be severely impacted by hybridization, which would degrade the genetic integrity of this already endangered species.

Today, thanks in large part to conservation efforts in the early twentieth century, the American alligator survives across much of the southeastern region of North America (Figure 1). The northern limits of its range are determined by low winter temperatures, and thus it is restricted to habitats south of Virginia along the Atlantic Coastal Plain. In the Mississippi River Valley, its northern limit is extreme southwest Tennessee, extreme southeast Oklahoma, and south central Arkansas. In 1979, alligators were released in the artificial lakes of the impounded Tennessee River in north Alabama, and a small population still survives there. Increased sightings of alligators at other sites within the state of Tennessee suggest that they are expanding their range, possibly in response to current climate change.

In addition to the living species of crocodilians, paleontologists have identified many species of extinct crocodilians from fossils, some of which were quite enormous—even gigantic—in size, with skulls up to 4 to 6 feet long (1 to 2 meters) and estimated total lengths up to 40 feet (12 meters). The saltwater crocodile (*Crocodylus porosus*) is considered the largest living crocodilian, with a total length up to 23 feet (7 meters), although larger individuals have been reported in the past. However, maximum lengths of the living crocodilians are controversial since many authorities consider the older records exaggerated.

3. Michael R. Rochford, et al. 2016, "Molecular Analyses Confirming the Introduction of Nile Crocodiles, *Crocodylus niloticus* Laurenti 1768 (Crocodylidae), in Southern Florida, with an Assessment of Potential for Establishment, Spread, and Impacts," *Herpetological Conservation and Biology* 11, no. 1 (2016): 80–89.

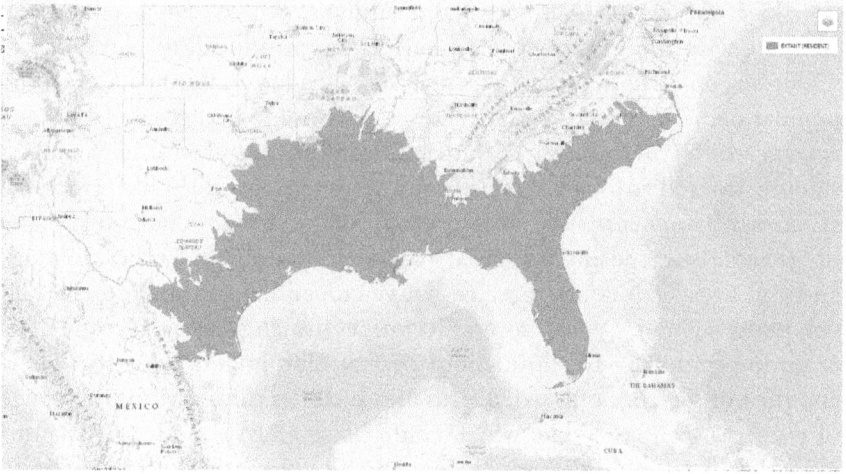

Figure 1. Distribution of the American Alligator, 2018
Source: Elsey, R. & Woodward, A. *Alligator mississippiensis*.
The IUCN Red List of Threatened Species, 2018.

The maximum size of the American alligator is also controversial. Early explorers and naturalists in North America during the 1700s reported sizes exceeding 20 feet (6 meters), but the largest recorded sizes for alligators since 1900 have been less than 16 feet (4.9 meters). This contrast has caused many alligator authorities to challenge the larger lengths as exaggerations, although it is feasible that exceptionally large individuals could have survived prior to the wholesale slaughter for hides begun in the mid-1800s. A 2015 study emphasized the need to use standardized techniques for measuring and documenting record-sized alligators and now recognize 14 feet, 9.25 inches (4.5 meters) to be the longest "credible record" for the American alligator.[4]

The largest record considered to be valid by some alligator authorities was of a 19-foot, 2-inch (5.84 meter) alligator killed and measured in 1890 by Edward McIlhenny near Vermilion Bay, Louisiana.[5] However, even this record has been challenged as being unverified.[6] McIlhenny himself admitted that the longer alligators were exceptional and that 15 feet (4.5 meters) is the

---

4.   Based upon straight-line total length measurement. A. Brunell, et al., "A New Record for the Maximum Length of the American Alligator," *Southeastern Naturalist* 14, no. 3 (2015): N38.

5.   E. A. McIlhenny, *The Alligator's Life History* (Christopher Publishing House, 1935).

6.   Brunell, et al., "A New Record"; Allan R. Woodward, et al., "Maximum Size of the Alligator (*Alligator mississippiensis*)," *Journal of Herpetology* 29, no. 4 (1995): 507–13.

normal maximum size for adult male alligators. Females are smaller, with a maximum size of about 9 feet (2.7 meters). McIlhenny suggested that longer lengths would probably never again be attained because of the intense hunting pressure for skins.

McIlhenny recorded his encounter with this, the largest alligator he had ever seen, in his excellent book on the life history of alligators. While on a duck hunt along the edge of Vermilion Bay at dusk, he killed the alligator with a shot to the head and then returned the next morning to measure it. Using his gun barrel, which he knew to be 30 inches long, he carefully measured the alligator's length three times, yielding the record length of 19 feet, 2 inches. He noted that the badly discolored teeth were worn down almost to the jawbone, suggesting a very old animal. McIlhenny also reported several other alligators killed in the late 1800s near Avery Island that measured over 17 feet long (5.2 meters). One of these was familiar enough to the family to be recognized as "Old Monsurat," killed in 1879 by McIlhenny's tutor, Robert Moony, and measured at 18 feet, 3 inches (5.56 meters) by McIlhenny's father, who was quoted as saying it was the largest alligator he had ever seen. In view of McIlhenny's recognized expertise, there is little reason to doubt the validity of these records.

Chapter Three

Prehistoric Alligator–Human Interactions

According to paleontologists, American alligators have existed in southeastern North America for an estimated eight million years at least. Fossils have been found from Florida to Texas that are virtually identical to the modern species, along with several extinct crocodilian species. The alligator likely had few competitors or predators as adults, except for the other species of crocodilians, until prehistoric humans migrated into the region some 12,000 to 15,000 years ago. Since these early peoples left no written records, one major source of information regarding their culture and how they interacted with alligators are the stone and bone tools left behind at the scattered camps they occupied. As Native peoples became more numerous and occupied sites for longer periods of time, they left more evidence of their former presence at kitchen middens (or dump sites for waste materials), including stone tools, animal bones, mollusk shells, and various artifacts. Eventually, as might be expected over more than 12,000 years, many changes occurred as the population grew, diversified into separate cultures, and evolved new techniques and skills. Additionally, they began to occupy home sites for longer periods of time.

This prehistoric period in North America, particularly for cultures in what is now the southeastern United States, is commonly divided into three major eras: the Paleo-Indian Era (15,000 to 6500 BC), the Archaic Era (6500 to 2000 BC), and the Woodlands and Mississippian Eras (2000 BC to 1500 AD).[1] These eras were followed by the historical period, when European explorers reached the continent and produced written records of their explorations.[2]

1. Some twentieth-century scholars referred to the Archaic Period as the Meso-Indian Era, and the combined Woodlands and Mississippian Periods as the Neo-Indian Era.
2. Kathleen M. Byrd and Robert W. Neuman, "Archaeological Data Relative to Prehistoric Subsistence in the Lower Mississippi River Alluvial Valley," in *Man and Environment in the Lower Mississippi Valley*, ed. Sam B. Hilliard (Louisiana State University, 1978); W. G. Haag, "A Prehistory of the Lower Mississippi River Valley," in *Man and Environment in the Lower Mississippi Valley*, ed. Sam B. Hilliard (Louisiana State University, 1978); Robert W. Neuman and Nancy W. Hawkins, *Louisiana Prehistory (Anthropological Study No. 6)* (Louisiana

There is very little evidence that Paleo-Indians had any interaction with alligators, although alligators could have been an important source of food, as they were for later cultures. The population density of humans at the time was quite low, with most people living in small, nomadic groups, and few sites indicate long-term occupation.[3] Archaeological excavations suggest that these sites were primarily temporary hunting camps, with few artifacts other than stone tools used for killing and butchering prey. As opportunistic hunters, they most likely considered any animal, including alligators, as potential food.

Some authorities have suggested that excessive hunting by the Paleo-Indian immigrants to North America could have been a significant force in the extinction of several species, such as Pleistocene megafauna including mammoths and mastodons, which disappeared after about 9000 BC, although most archaeologists now believe that climate change was a more significant factor.[4] The rate at which alligators were killed by sparsely populated prehistoric hunters was probably never sufficient to seriously impact their population numbers.

As the human population continued to increase in the Archaic Era, people began to remain in one location for longer periods of time before moving on, leading to larger middens with increased evidence of site occupation. In particular, they left behind increased volumes of food wastes, such as clam and oyster shells, bones (including alligator bones in some), and other wastes. More permanent sites provide greater evidence of the cultures of these Archaic peoples.

During the Woodlands and Mississippian Periods, human cultures experienced important new developments, including improvements to their hunting techniques. While the Paleo-Indians used simple spears or javelins, Archaic Indian cultures developed the atlatl (or spear-thrower) to increase the force and distance of their spears. The Woodlands and Mississippian groups improved upon this system yet again with the creation of the bow and arrow.[5] Such advances made the hunter more effective in killing alligators and other large prey. Moreover, these later cultures developed agriculture, which required them to settle in an area long enough for crops to grow and be harvested. This led to a proliferation of artifacts and remains in middens.

Department of Culture, Recreation and Tourism, Louisiana Archaeological Survey and Antiquities Commission, 1987).

3. Neuman and Hawkins, *Louisiana Prehistory*; M. A. Rees, *Archaeology of Louisiana* (Louisiana State University Press, 2010).

4. Lynda N. Shaffer, *Native Americans Before 1492: The Moundbuilding Centers of the Eastern Woodlands* (Routledge, 1992).

5. Byrd and Neuman, "Archaeological Data Relative to Prehistoric Subsistence."

Archaeological excavations of shell middens in the southeast coastal plain demonstrate that alligators were a common food for many in the Archaic and Woodlands/Mississippian periods, but much less important to their diet than larger mammals at inland sites (especially white-tailed deer, bears, and bison) and marine species (fish and shellfish) along the coast. However, some studies of shell mounds in coastal Louisiana have demonstrated that, for the people in this region, alligators were second only to deer as an important source of animal protein based upon weight and nutritive value.[6]

Though alligator bones were abundant at many sites, especially in Louisiana and Florida, several of the middens excavated within the natural range of alligators failed to yield alligator bones, suggesting that their importance as food may have varied significantly among different groups. This absence of bones at archaeological sites does not prove that Indigenous peoples did not eat alligators at those sites, but it does suggest that alligators were not a dominant part of the local diet. An important example is the Bottle Creek site in the Mobile-Tensaw Delta of Alabama, where alligators are abundant and could have been a significant, reliable, and rather easily obtained source of protein. Instead, nearby estuarine habitats yielded the fish and mollusks that served as the dominant foods.[7] Other prehistoric groups appear to have avoided or rejected alligators completely. For example, John R. Swanton, noted authority on North American Indigenous peoples, was convinced that Natchez people did not eat alligators (see Chapter 4). Of course, possible local cultural preferences for certain foods may be a determining factor in such differences.

Alligators as Religious or Totemic Symbols

Among the artifacts left behind by prehistoric Indigenous peoples are what appear to be religious or spiritual relics, some of which seem to involve alligators. Rather than avoiding or tolerating the species, these groups respected and honored alligators as a unique and special animal. In several

6. Kathleen M. Byrd, "Tchefuncte Subsistence: Information Obtained from the Excavation of the Morton Shell Mound, Iberia Parish, Louisiana," *Southeastern Archaeological Conference Bulletin* 19 (1976): 70–75; Kathleen M. Byrd, "The Brackish Water Clam (Rangia cuneata): A Prehistoric 'Staff of Life' or a Minor Food Resource," *Louisiana Archaeology* 3 (1976): 23–31; William G. McIntire, "Prehistoric Settlements of Coastal Louisiana" (PhD Diss., Louisiana State University, 1954), https://digitalcommons.lsu.edu/gradschool_disstheses/8099.
7. Irvy R. Quitmyer, "Zooarchaeological Remains from Bottle Creek," in *Bottle Creek: A Pensacola Culture Site in South Alabama*, ed. Ian W. Brown (University of Alabama Press, 2003).

cases, the artifacts suggest alligators had godlike attributes. However, the actual importance of such beliefs is difficult to assess without written records. We must depend upon archaeological evidence and oral traditions passed down to their descendants.

Many Native American cultures had alligator dances, which may have had religious and possible clan or totemic significance. An especially impressive example of such possible dances is represented by artifacts excavated by anthropologist Frank Hamilton Cushing in 1896 at a Calusa village site on Key Marco, Florida.[8] These impressive items included carved and painted wooden heads of animals, including an alligator, and matching masks, estimated to have been carved 500 to 1500 years before they were excavated. Although they have never been accurately dated, they are certainly pre-Columbian, possibly as old as one thousand years before the invasion of Europeans, although scholar Marion Gilliland suggested a late fifteenth-century date. Researchers believe masked dancers may have carried these carved heads to represent the animal and its clan affiliation.[9] Such wooden artifacts typically deteriorate over time and as such are rarely preserved, but these survived for so long because they were covered by oxygen-free muck that prevented their decay. Unfortunately, many of these items began to deteriorate after discovery because they were not protected while drying. Fortunately, crew member Wells M. Sawyer recorded as many as he could in photographs and watercolors as they were removed from the muck. One of the most celebrated of these paintings featured the alligator head, illustrated in color in Jerald T. Milanich's book, *Florida's Indians: From Ancient Times to the Present*.[10] Other artifacts from the Key Marco site illustrated by Gilliland include alligator bone points, which possibly functioned as spear points, and a wooden box lid and sides decorated with detailed drawings of alligator-like animals with a horn that suggests a mystical creature.[11] The surviving alligator head and other artifacts from this location are now protected and preserved in museums, including the Smithsonian

8.   Emma Lila Fundaburk, et al., *Sun Circles and Human Hands: The Southeastern Indians Art and Industries* (University of Alabama Press, 2001); Marion S. Gilliland, *The Material Culture of Key Marco, Florida* (University Press of Florida, 1975); Kelby Ouchley, *American Alligator: Ancient Predator in the Modern World* (University Press of Florida, 2013); Lucy Fowler Williams, "The Calusa Indians: Maritime Peoples of Florida in the Age of Columbus," *Expedition* 33, no. 2 (1991): 55–60.
9.   Ouchley, *American Alligator*; Williams, "The Calusa Indians."
10. Jerald T. Milanich, *Florida's Indians: From Ancient Times to the Present* (University Press of Florida, 1998), Plate 9.
11. Gilliland, *The Material Culture of Key Marco*.

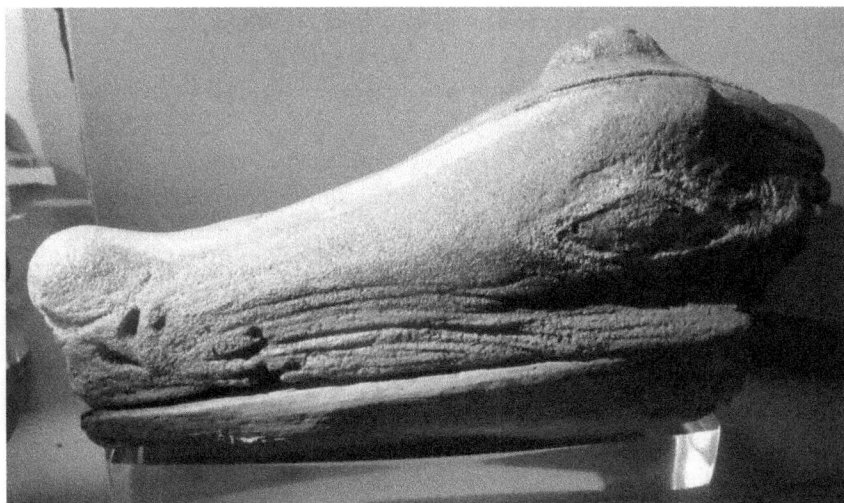

Carved and painted alligator head
(replica)
Wood and paint
A.D. 700-1500
Key Marco, Collier Co.
Original at University of Pennsylvania
Museum of Archaeology and Anthropology

(*Above*) Replica of carved wooden figurehead of alligator,
original excavated from Key Marco in 1896, with (*below*) adjacent museum description.
Photographed in Florida Museum of Natural History.

National Museum of Natural History and the University of Pennsylvania
Museum in Philadelphia. (In addition, replicas are displayed at several other
museums, including the Florida Museum of Natural History.)

Another significant development for later cultures was the beginning
of mound building, which often resulted in massive shell or earthen structures. Some were burial mounds, where the remains of people were buried along with various artifacts, sometimes including other animals, and at
least one site that included a ceremonial alligator burial. Other flat-topped

mounds served as ceremonial sites or locations for lodges, temples, or other important structures.[12]

In addition, Woodlands and Mississippian cultures sometimes formed the constructed mounds in the shape of alligators or other animals, today referred to as effigy mounds. The function of such mounds is unknown but apparently had religious ritual or ceremonial significance related to alligators. Unfortunately, two of the more significant of these in coastal Louisiana, the Alligator Effigy Mound on the southeast shore of Grand Lake and the Morton Shell Mound on the northwest side of Weeks Island, have been mostly destroyed by shell dredging and erosion from wave action and sea level rise.[13]

One of the most intriguing burial mounds is the Palmer Burial Mound, which includes the ceremonial burial of an alligator. This site is located at Spanish Point (now the location of Selby Gardens) in Osprey, Florida, one of the most significant archaeological preserves in Florida. The site was occupied by Native Americans from about 2100 BC to 1100 AD.[14] In 1910, Ms. Bertha Palmer purchased the property, and she is primarily responsible for preserving the site, now known to archaeologists as the Palmer Site. This area includes three shell middens created during sequential periods of occupation: the Archaic or Hill Cottage Midden (2500 to 1000 BC); the Shell Ridge Midden (before 300 BC to after 150 AD); and the Southeast Shell Midden and adjacent Burial Mound (150 AD to 1100 AD).[15]

A mystery regarding the Palmer Site is that, in addition to the more than four hundred human burials that were excavated at the Burial Mound, archaeologists found four dogs and one "ceremonial interment of an alligator."[16] Two strings of sawfish (*Pristis* sp.) vertebrae beads were placed parallel to the alligator body, one 5 feet long (1.52 meters), stretching from forelimb

---

12. Shaffer, *Native Americans Before 1492.*

13. Trent Gremillion, "Alligator Effigy Mound," in *Southwest Louisiana Archaeology* vol. 1 (Kirkman House, 2019); H. V. Howe, et al., *Bulletin 6: Reports on the Geology of Cameron and Vermilion Parishes* (Dept. of Conservation, Louisiana Geological Survey, 1935).

14. Ripley P. Bullen, "Some Variations in Settlement Patterns in Peninsular Florida," *Bulletin No. 13: Proceedings of the Twenty-Seventh Southeastern Archaeological Conference,* ed. Bettye J. Broyles (1971), 10–19; Ripley P. Bullen and Adelaide K. Bullen, *The Palmer Site,* Florida Anthropological Society Publications no. 8 (Florida Anthropological Society, 1976); Dale L. Hutchinson, *Bioarchaeology of the Florida Gulf Coast: Adaptation, Conflict, and Change* (University Press of Florida, 2004); Michael Russo and Irvy R. Quitmyer, "Developing Models of Settlement for the Florida Gulf Coast," in *Case Studies in Environmental Archaeology. Interdisciplinary Contributions to Archaeology,* eds. E. J. Reitz, S. J. Scudder, and C. M. Scarry, (Springer, 2008).

15. Bullen and Bullen, *The Palmer Site.*

16. Bullen and Bullen, *The Palmer Site,* 44.

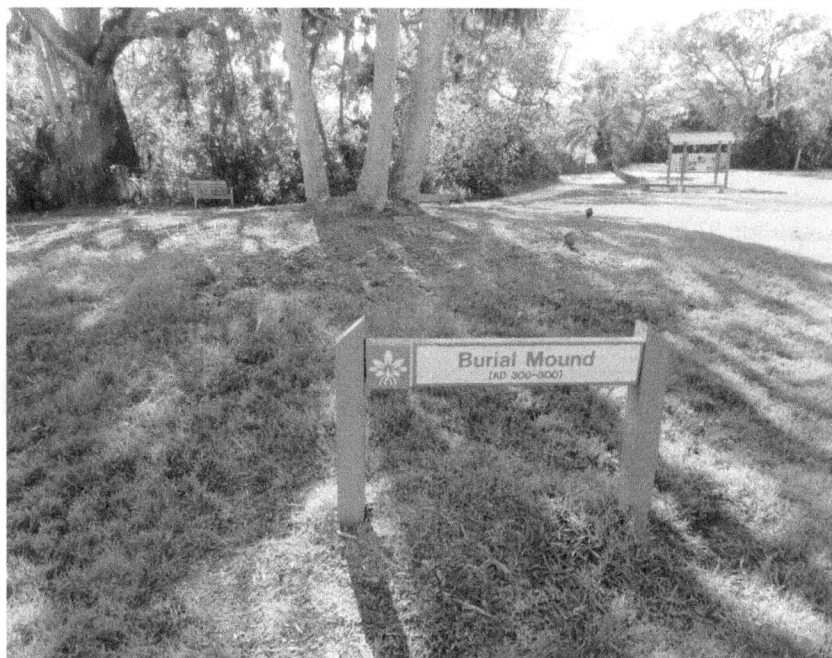

Palmer Burial Mound at Selby Gardens, Osprey, Florida, February 9, 2023

to hindlimb, and closely associated with the body; the other was 3 feet long (0.9 meters) and placed 6 to 12 inches (15 to 30 centimeters) away and 5 inches (13 centimeters) above. Archaeologists Ripley and Adelaide Bullen suggested that this placement implies that the alligator "was laid in a shallow pit [11 inches (28 centimeters) deep from the 1959 mound surface] and the shorter string deposited on the side of the pit."[17] The size of the alligator was not recorded but assuming that the distance between the limbs is accurate, the alligator must have been at least 15 feet (4.6 meters) long. It is not known why the alligator was buried at this site, though it is clear from the unique placement of the animal that the act was ceremonial and "a much more important event than the dog burials."[18] Bullen and Bullen suggested two possibilities. The first, that the alligator could have eaten an important person, was dismissed as no human bones were found in close association with the alligator. They considered more likely the possibility that this burial implied

17. Bullen and Bullen, *The Palmer Site*, 46.
18. Bullen and Bullen, *The Palmer Site*, 46.

"some type of totemic relationship."[19] Also significant: the alligator is thought to be the last burial in the Palmer Burial Mound. It may be that the last inhabitants of the Palmer Site were members of an alligator clan.

There was no evidence that alligators were eaten by any of the Palmer Site residents. Instead, their primary food sources were marine shellfish, dominated by oysters and scallops but also including numerous other mollusks and fish.[20] Turtles, deer, and other animals rounded out their diets, though in much smaller quantities.

Another important burial mound site is the Lake Jackson Mounds Archaeological State Park near Tallahassee, Florida. Milanich included alligators in a list of twenty animals represented at this site and suggested that all of these animals most likely were represented in various myths and ceremonies by the people who occupied this site from about 1000 to 1500 AD.[21]

The various tools or decorations found in these archaeological excavations, such as awls, needles, pins, chisels, hoes, fishhooks, projectile points, and pendants, were often made from bones, including those of alligators, although most would have been made from the much more common deer bones. Alligator teeth were among the most common artifacts found, reflecting their value as trade objects, and were often found outside the normal range of alligators, indicating they were traded for other goods.

The spiritual or religious significance of alligators to some Native American cultures is not surprising in view of the numerous examples of crocodilian worship in both Old World and New World cultures. For example, the Egyptian crocodile cult dating back at least to 1800 BC involved worship of the crocodile-headed god Sebek (or Sobek), and archaeologists have found evidence of crocodile breeding locations similar to modern alligator rearing facilities. Sometimes the ancient Egyptians even embalmed, mummified, and buried crocodiles in coffins.[22]

Similarly, crocodile or caiman worship features prominently in the Olmec culture of Central America dating back to about 1200 BC (both preceding and influencing the later Aztec and Mayan civilizations). In their 1980 article, Terry Stocker, Sarah Meltzoff, and Steve Armsey asserted that "the crocodilian may have attained religious importance because it served as a

19. Bullen and Bullen, *The Palmer Site*, 46.
20. Bullen and Bullen, *The Palmer Site*; Russo and Quitmyer, "Developing Models of Settlement."
21. Milanich, *Florida's Indians*.
22. David Alderton, *Crocodiles & Alligators of the World* (Facts on File, 1991); Zach Fitzner, "In a Number of Different Cultures, Crocodiles Are Worshipped," Earth.com, March 28, 2019, https://www.earth.com/news/cultures-crocodiles-worshipped/.

major food source, is a predator of humans, and is an anomalous and striking animal."[23] They further state that "crocodilian products (meat, skins, etc.) may have been" used in trade with other groups.

The Olmec Dragon, described as a crocodile-like god, represented the earth and was possibly the predecessor of the Aztec god Cipactl (which means "crocodile"). As noted by David Alderton, "Why such similar stories have arisen about crocodilians throughout the world, amongst isolated tribes and before the days of European influence, is a mystery."[24] How (and if) these impressive Central American civilizations and their beliefs could have affected North American prehistoric peoples is unknown. However, there are numerous examples of human-alligator connections that preceded the arrival of Europeans.

23. Terry Stocker, et al., "Crocodilians and Olmecs: Further Interpretations in Formative Period Iconography," *American Antiquity* 45, no. 4 (1980): 740.
24. Alderton, *Crocodiles & Alligators*, 740.

Chapter Four

Historical Accounts of Alligator-Human Interactions

A s European explorers and settlers began invading southeast North America in the 1500s, they documented the Indigenous cultures they encountered, basing much of what they wrote on direct observations as well as the oral traditions of the Native Americans—of course, only after establishing a means of communication. Such European accounts have, in many ways, enhanced our understanding of prehistoric Native cultures, but also highlight the biases and perceptions of the authors.

Moreover, as time passed and interactions between such diverse cultures increased, European immigrants began to influence and modify Native cultures, intentionally or otherwise. For example, Europeans brought previously unknown domesticated plants and animals, iron tools and other implements including firearms, and diseases. They also brought new religions and pushed conversion on many Native groups. Thus, over the next several hundred years, the arrival and dominance of Europeans had a profound influence on Indigenous cultures. Similarly, though to a lesser degree, Indigenous Americans influenced the habits and attitudes of the European immigrants, such as their adoption of available local foods like maize, turkey, and potatoes. In some cases, the Native peoples also inspired changes regarding European attitudes toward alligators as food.

Alligators as Food

Many European writers recorded incidents involving alligators in which they themselves consumed alligator meat, sometimes to avoid starving. Few European commentators appeared to enjoy alligator meat, although French explorer Pierre Le Moyne, Sieur d'Iberville, in 1699 described them as "very good to eat, the flesh being very white and delicate but smelling of musk, which is a scent that the flesh must be rid of before one can eat it."[1] Nicolas Le Challeux, a member of the French Huguenot colony that erected the ill-fated

---

1. R. G. McWilliams, trans. and ed., *Iberville's Gulf Journals* (University of Alabama Press, 1981), 82.

Fort Caroline in 1564 at the mouth of the St. Johns River, also wrote of the palatability of alligator meat. In his *Discours de l'Histoire de la Floride*, he reported on a meal of "crocodilles" (alligator) in which the meat was "tender & white like that of a calf [veal], & about the same taste."[2] In contrast, his colleague René Goulaine de Laudonnière wrote that alligator flesh "is fair and white, and, were it not that it savored too much like musk, we would oftentimes have eaten thereof."[3] However, both Le Challeux and Laudonnière wrote of the starvation that the French colony experienced and that they repeatedly begged for food such as maize, beans, and fish from the local Timucua people that befriended them, or they were forced to eat herbs, roots, and other foods they could scavenge. In spite of their hunger, Laudonnière never mentions anyone in the colony eating alligator, even though alligators were abundant in the area and a common food of the Timucua. It seems the French colonists did not know the recommended methods of avoiding the musk glands when butchering an alligator, leading to the unappetizing smell. Moreover, they lacked the knowledge to avoid fat, select the best meats such as the tail, and choose alligators less than six feet long.[4]

While traveling down the Mississippi River with French explorer René-Robert Cavelier, Sieur de La Salle, in 1682, Nicolas de La Salle (not related to the explorer) reported that the group killed an alligator, providing a "feast . . . which was quite good."[5] He estimated this event took place approximately thirty leagues (ninety miles) below the mouth of the Arkansas River (or just downriver from the current site of Vicksburg). Two or three days later, the expedition visited an Indigenous village located on Lake St. Joseph (in Tensas Parish, Louisiana), where the Taensa people provided them with various provisions, including figures of alligators and other animals made of a paste

2. Nicolas Le Challeux, *Discours de L'Historie de la Floride Contenant.la Cruaute des Espangnola, Contre les Sujects du Roy* (n.p.: De Dieppe, 1566), 21, https://archive.org/details/discoursdelhisto00lech/page/n7/mode/2up.
3. René Goulaine de Laudonnière, "History of the First Attempt of the French (The Huguenots) to Colonize the Newly Discovered Country of Florida," in *Historical Collections of Louisiana and Florida, including Translations of Original Manuscripts Relating to Their Discovery and Settlement, with Numerous Historical and Biographical Notes*, ed. B. F. French (New York: J. Sabin & Sons, 1869), 175, online facsimile edition at www.americanjourneys.org/aj-141/. Accessed November 14, 2024.
4. Vaughn L. Glasgow, *A Social History of the American Alligator: The Earth Trembles with His Thunder* (St. Martin's Press, 1991).
5. William C. Foster, *The La Salle Expedition on the Mississippi River: A Lost Manuscript of Nicolas de La Salle, 1682*, ed. William C. Foster, trans. Johanna S. Warren (Texas State Historical Association, 2003), 106.

composed of fruit and corn flour.[6] The Taensa also warned the explorers that downriver they would find vicious tribes who would cannibalize them.

Zenobius Membré, also traveling with La Salle, noted that while exploring the lower Mississippi River the group ran out of provisions and subsequently lived only on potatoes and alligators.[7] Membré further wrote that near the mouth of the river they found some dried meat, which they took "to appease our hunger; but soon perceiving it to be human flesh, we left the rest to our Indians. It was very good and delicate."[8] Other chroniclers of this expedition provided additional details of this event but with some variation. Nicolas de La Salle wrote that they "came upon a canoe with three Indians"[9] who ran off, leaving in the canoe, some smoked alligator, and another piece of meat. After eating it all, the explorers "realized later by the bones, that the meat was from human top ribs. The human meat was better than that of the alligator."[10] Minet, an engineer on the voyage who kept a journal, recounts of the same event, "As hunger was pressing us, having only a little corn daily, we pounced on this meat. When we had eaten it, we knew that it was human by the bones and the taste, which was better than the caiman [alligator]."[11]

Just three days previously, on April 14, near present-day New Orleans, they had seen Indigenous people fishing on the right bank of the river. These people also ran away, leaving behind a basket containing a fish, a man's foot, and a child's hand, all smoke-dried.[12] This "vicious tribe," of whom the Taensa had warned La Salle, were likely the Quinapisa (sometimes spelled Quinapissa) living on the west bank of the Mississippi.[13]

In 1591, Theodor de Bry published a collection of engravings that he claimed were based on sketches by Jacques Le Moyne de Morgues—artist for a 1564 French expedition to Florida—as well as accounts from several European explorers. One piece shows Timucua people smoking alligators and

6. Foster, *The La Salle Expedition*; Robert S. Weddle, et al., eds., *La Salle, the Mississippi, and the Gulf: Three Primary Documents* (Texas A&M University Press, 1987).
7. John G. Shea, *Discovery and Exploration of the Mississippi Valley with the Original Narratives of Marquette, Allouez, Membré, Hennepin, and Anastase Douay* (Joseph McDonough, 1903).
8. Shea, *Discovery and Exploration*, 175.
9. Foster, *The La Salle Expedition*, 114.
10. Foster, *The La Salle Expedition*, 114.
11. Weddle, et al., *La Salle, the Mississippi*, 54.
12. Foster, *The La Salle Expedition*.
13. Foster, *The La Salle Expedition*, 107, 113; Paul Chesnel, *History of Cavelier De La Salle 1643–1687 Explorations in the Valleys of the Ohio, Illinois and Mississippi*, trans. Andrée Chesnel Meany (G. P. Putnam's Sons, 1932), 146.

Timucua of north Florida drying alligators and other animals for preservation. *Source*: Florida Memory, State Library and Archives of Florida.

other animals to dry them for preservation.[14] This and other works are the earliest known European images of Native Americans in Florida, but the accuracy of details in these figures has been questioned by some archaeologists.[15]

Anthropologist John Reed Swanton provided another example of a Native American nation, the Natchez, who apparently did not eat alligators. He noted that ethnographer and scholar Antoine Simon Le Page du Pratz, who spent eight years living with the Natchez, made no mention of alligator as a food animal for the group, which "is noticeable, as it was much esteemed along most of the Gulf coast" and was an important food source, especially for tribes in the Florida peninsula and the coast tribes of Louisiana.[16] According to Le Page du Pratz, the Natchez chiefly ate meats from "the buffalo [bison], the deer, the bear, and the dog,"[17] whereas Swanton wrote that the nations that Le Page du Pratz mentioned as eating alligators "were probably those

14.   Michael Alexander, ed., *Discovering the New World, based on the works of Theodor de Bry* (Harper and Row, 1976), 41, 43; engravings also available at State Archives of Florida, Florida Memory, https://www.floridamemory.com/.
15.   Jerald Milanich, "Fact or Fiction: Theodore De Bry's 1591 Engravings of Early Florida Indians," *Adventures in Florida Archaeology* (2016): 21–28. According to Milanich, a noted archaeologist from Florida, "Bry's images of Florida Indians are bogus. They are not based on paintings done by Jacques Le Moyne. Everything depicted in them . . . should be questioned." Milanich, "Fact or Fiction," 22.
16.   John R. Swanton, *The Indians of the Southeastern United States*, Bureau of American Ethnology Bulletin 137 (US Government Printing Office, 1946), 291.
17.   Antoine Simon Le Page du Pratz, *The History of Louisiana, or of the Western Parts of Virginia and Carolina: Containing a Description of the Countries that Lie on Both Sides of the River Mississippi: With an Account of the Settlements, Inhabitants, Soil, Climate, and Products* (Paris: n.p., 1774; facsimile e., Louisiana State University Press, 1975), 349.

of southwestern Louisiana, not the Natchez."[18] Supporting the concept that the Natchez (and possibly other nations) did not commonly eat alligator, Swanton references what he calls the "Luxembourg memoir," an anonymous work entitled *Mémoire sur La Louisiane on Le Mississipi*, published in Luxembourg in 1752. Written primarily about Natchez in the early 1700s, this document states that people who have lived badly were expected after death to revive in "a miserable nation and in a country where only alligators are eaten."[19] This difference may reflect the relative availability of alligators compared to other animals and their ease of capture, as well as a possible preferences for other foods.

The Alligator and Religion

The alligator played a significant role in the religious and cultural practices of Native groups in much of North America. In many Indigenous nations, including the Alabama, Caddo, Chickasaw, Creek (or Muskogean), and Natchez, the species must have held cultural importance, as "alligator" was one of the names used by clans within the larger group. Moreover, the alligator was an important totem figure for the Bayogoula, Choctaw, Chickasaw, and others.[20] William Bartram, an American naturalist who traveled through the Carolinas, Florida, and Georgia in the late 1700s, informed Swanton that the Creek used the alligator as the symbol of their town of Tukabatchee in modern-day Alabama, where the front pillars in the central square grounds were carved like alligators.[21]

Numerous European writers described the spiritual beliefs of the Native American groups they encountered, many of which included the crocodilians. However, we should always remember that commentary on spiritual beliefs can be quite complex and mysterious, especially given European biases and language barriers. Dumont, writing about the "Tenças" (perhaps referring to the Taensa), residents near the Natchez, stated that they "worship the god they choose themselves. Some worship the sun, others the moon, some the snake, the alligator, etc., according to their wishes."[22]

18. John Reed Swanton, *Indian Tribes of the Lower Mississippi Valley and Adjacent Coast of the Gulf Of Mexico*, Bureau of American Ethnology Bulletin 43 (US Government Printing Office, 1911), 72.
19. Swanton, *Indian Tribes of the Lower Mississippi*, 181.
20. Swanton, *Indian Tribes of the Lower Mississippi*; Swanton, *The Indians of the Southeastern United States*.
21. Swanton, *The Indians of the Southeastern United States*, 617.
22. Dumont de Montigny, *The Memoir of Lieutenant Dumont*, 345.

In both Old and New World cultures, crocodilian teeth and other parts were thought to have mystical powers, such as preventing or curing diseases, and were often worn around the neck as a talisman.[23] Such false concepts of crocodilian power have survived even up until relatively recent times. University of Florida biologist Kent Vliet wrote of an incident in 1820 near Amarillo, Texas, where members of the Stephen H. Long expedition observed a band of Kiowa-Apaches—who lived outside the range of alligators—wearing ornaments depicting alligators.[24] These included amulets carved of wood wrapped in leather and decorated with white and blue beads, apparently worn to prevent or cure diseases. The account claims the chief pressed a small wooden alligator repeatedly against an open wound.

Several nations had alligator dances and songs with apparent religious and clan significance (as well as use in medical treatments). In 1700, Jesuit missionary Father Jacques Gravier described a treatment that he observed by Houma medicine men in which "one whistled and played on a gourd, another sucked [blood from scarified places on the body], while the third sang the song of the alligator, whose skin served him as a drum."[25] Swanton also included alligators in a list of animals that were thought by various Native American groups of the Southeast to cause diseases.[26] Swanton quoted an unidentified and previously unpublished book, *Relation de la Louisiane* (found in the Edward E. Ayer Collection at the Newberry Library in Chicago), which he tentatively dated as 1755 or earlier, as follows: "They [Choctaws] have dances among them accompanied by feasts, which are almost alike. Only the names differ, as the dance of the turkey, bison, bear, alligator. In this last they have masks made like the head of this animal, one or two disguising themselves thus, while five or six others take masks of different animals which the alligator commonly eats, and then they make a thousand grotesque antics."[27] Swanton suggested that the author of *Relation* seemed to have been a prominent French officer.

23. Alderton, *Crocodiles & Alligators*.
24. Vliet, *Alligators*.
25. Jacques Gravier, "Journal of the Voyage of Father Gravier," in *Jesuit Relations and Allied Documents: Travels and Exploration of the Jesuit Missionaries in New France, 1610–1791*, vol. 66, ed. Rueben G. Thwaites (Cleveland: Burron Brothers Company, 1900); Swanton, *Indian Tribes of the Lower Mississippi*, 289; Bryan L. Guevin, "The Ethno-Archaeology of the Houma Indians" (PhD diss., Louisiana State University, 1983), 44, https://digital commons.lsu.edu/gradschool_disstheses/8310.
26. Swanton, *The Indians of the Southeastern United States*, 794.
27. John Reed Swanton, *Source Material for the Social and Ceremonial Life of the Choctaw Indians*, Bureau of American Ethnology Bulletin 103 (US Government Printing Office, 1931), 221.

Alligators served as the basis for various stories, myths, and legends in Native American cultures. In 1929, Swanton published a list of 305 myths and tales of the southeastern nations (including the Creek, Hitchiti, Alabama, Koasati [Coushatta], and Natchez peoples), at least eight of which included alligators.[28] Elsbeth Dowd, writing of Caddo culture in *Southeastern Archaeology*, noted that stories involving alligators followed a common theme of "power," and even though they are viewed as powerful and potentially dangerous, they could also be helpful, giving people special power.[29]

An extreme case of reverence for alligators (as well as snakes) was recorded in the writings of Dutch navigator Bernard Romans in 1775. In reference to the "Arkanzas" nation (likely the Quapaw) occupying the area around Arkansas Post near the junction of the Arkansas, White, and Mississippi Rivers, Romans stated that they "have so far a veneration for the alligator as not to destroy him, nor have I seen a savage who would willingly kill a snake."[30]

In addition to the religious, medical, and ceremonial significance, historical and modern writers have described other functions that alligators have served for Native Americans. Surprisingly there is little evidence that Indigenous peoples used alligator skins to any great extent as clothing, although they would sometimes clothe themselves in such skins as costumes in order to represent evil spirits and scare others.[31] According to a 1915 dictionary of the Choctaw language, a *tilikpi* is "an ancient kind of shield, made of stiff hide of a cow, or of an alligator."[32] The stiff, heavy armor on the dorsal surface of an alligator hide, with its bony osteoderms (bony plates), could have made an effective shield, just as it provided protective armor for the living alligator. In addition, a type of musical or rhythm instrument known as a rasp was played by rubbing a polished stick across a dried alligator skin, the equivalent of the modern-day washboard.[33] Preparing the alligator skins for this purpose involved "first exposing the alligator to ants until all of the softer parts had been eaten out and then drying the skin."[34] The Atakapa-Ishak of

28. John Reed Swanton, *Myths and Tales of the Southeastern Indians*, Bureau of American Ethnology Bulletin 88 (US Government Printing Office, 1929).
29. Elsbeth L. Dowd, "Amphibian and Reptilian Imagery in Caddo Art," *Southeastern Archaeology* 30, no. 1 (2015).
30. Bernard Romans, *A Concise Natural History of East and West Florida* (New York: R. Aitkin, 1775; facsimile ed., University of Alabama Press, 1999), 149.
31. Swanton, *Indian Tribes of the Lower Mississippi Valley.*
32. Cyrus Byington, *A Dictionary of the Choctaw Language* (US Government Printing Office, 1915), 351.
33. Swanton, *Source Material.*
34. Swanton, *Indian Tribes of the Lower Mississippi Valley,* 350.

southeast Texas and southwest Louisiana, who did eat alligator meat, considered alligator oil rendered from the animal's fat a delicacy, in addition to using the oil as an insect repellent and burning it in lamps.[35]

Fear of Alligators

Based on the written historical records, Native Americans seemed to have had little fear of alligators, in contrast to the Europeans, who regarded them as terrifying monsters. Many of the European explorers were familiar with Old World crocodiles and often referred to the alligators as caimans or crocodiles. European explorers also seemed to believe that the American alligators were just as ferocious and dangerous as some crocodiles. We may assume that some of the enslaved Africans forcibly transported to America would have brought with them their knowledge and opinions of African crocodiles and could have influenced the concepts that nearby Native groups had toward alligators.[36]

Paul du Ru, a Jesuit priest who accompanied Iberville in the early 1700s, noted that the Native people they encountered were not at all afraid of alligators but played with them while swimming.[37] Apparently even young Native American children were not afraid of alligators and often attacked them when they came on land. Le Page du Pratz wrote of an incident in the early 1700s when he retrieved his gun to shoot a 5-foot (1.5 meter) alligator but was stopped by a young Chitimacha girl, who proceeded to attack the alligator with a stick.[38] The following day he was told that it was common for children to pursue and kill young alligators on land. Le Page du Pratz also noted that boys and girls beginning as young as three years of age bathed every morning in nearby rivers, where they learned to swim and beat the water to "frighten away the crocodiles [sic]."[39] In 1722, French explorer and Jesuit Pierre François Xavier de Charlevoix was told that people bathing in the river had nothing to fear even though they were surrounded by alligators, that none of the alligators "came near them, and seemed only to watch them, in order to fall upon them, the moment they were going to leave the river; that then, in order to drive them away, they made a splashing in the water with a stick, which they took care to be provided with, and which made these animals

35. McIntire, "Prehistoric Settlements."
36. Ouchley, *American Alligator*, 69.
37. Paul du Ru, *Journal of Paul du Ru*, ed. and trans. Ruth L. Butler (Caxton Club, 1934).
38. Le Page du Pratz, *History of Louisiana*, 19.
39. Le Page du Pratz, *History of Louisiana*, 310.

fly to such a distance, that they had sufficient time to secure themselves."[40] Charlevoix might have the distinction of being the first European to promote the abuse of alligators, in that he described a lake in southeast Louisiana that could "furnish abundance of fish, were the alligators with which it swarms at present destroyed."[41] However, just fifty pages later he noted that the river near St. Marks, Florida, was "full of alligators, but for all that well stocked with fish,"[42] perhaps finally recognizing an important ecological principle of predator-prey relationships.

William Bartram confessed his own fear of alligators, which he described as "a very large and terrible creature, and of prodigious strength, activity and swiftness in the water."[43] Based upon one incident on the St. Johns River in Florida in 1774, when he was "pursued by several very large ones" and "attacked on all sides, several endeavouring to overset the canoe,"[44] it is not surprising that he might be somewhat fearful of alligators. However, he was able to fight off the alligators with a club and later suggested that this attack was by alligators "in search of fish," which he had in the boat.[45]

Later Europeans and European Americans seem to have become less fearful of alligators (possibly influenced by Native Americans they encountered). Artist and naturalist John J. Audubon described alligators as "truly gentle" and wrote of walking in waist-deep water with hundreds of alligators around him, and "thought nothing of it."[46] However, he considered them "dreadfully dangerous" in spring (the "love season") when the "heat of passion" and hunger made them ferocious and more active: "At this time, no man swims or wades among them."[47]

Edward McIlhenny described his childhood memories at Avery Island, Louisiana, of having no fear of alligators and swimming around

40. Pierre Francois Xavier de Charlevoix, *Journal of a Voyage to North America*, vol. II, trans. Louise P. Kellogg (London: R. and J. Dodsley, 1761; Caxton Club, 1923), 234.

41. Charlevoix, *Journal of a Voyage*, 266.

42. Charlevoix, *Journal of a Voyage*, 318.

43. William Bartram, *Travels Through North & South Carolina, Georgia, East & West Florida, the Cherokee Country, the Extensive Territories of the Muscogulges, or Creek Confederacy, and the Country of the Chactaws; Containing an Account of the Soil and Natural Productions of those Regions, Together With Observations on the Manners of the Indians* (Philadelphia: James & Johnson, Printer, 1791; repr., Dover Publications, 1928), 122.

44. Bartram, *Travels Through North & South Carolina*, 116.

45. Bartram, *Travels Through North & South Carolina*, 117.

46. J. J. Audubon, "Observations on the Natural History of the Alligator," *Edinburgh New Philosophical Journal*, vol. 2 (Oct. 1826–Apr. 1827): 6.

47. Audubon, "Observations," 7.

and under large alligators.[48] He and his friends would also abuse the alligators by "chunking" mud in their eyes and mouth, a risky behavior which is not recommended. He recalled that on one occasion, a large alligator, after being so abused, ran over Edward as it headed for the water and left scars on his stomach. This may have been the event that helped convert Edward from an abuser of alligators to a respecter and scientific observer of them.

Alligator Attacks on People

Though reports of the danger of American alligators are mostly exaggerated, these large, powerful carnivores are potentially dangerous and should be respected. Each year beginning around 1970 one or more alligator attacks on people have occurred, some of which were fatal. Ricky L. Langley, of the North Carolina Department of Health and Human Services, reported 376 bites between 1948 and 2004, including twenty-three deaths.[49] Florida recorded by far the largest number of attacks (334), followed by Texas (15), Georgia (9), South Carolina (9), Alabama (5), Louisiana (2), North Carolina (1), and Arkansas (1). Data from the Florida Fish and Wildlife Conservation Commission show that the number of bites in Florida increased substantially after about 1970 (Table 2) and have continued to increase through 2022, likely because of the rapid increase in the state's human population as well as the construction of housing developments, parks, and other attractions that place people in proximity to alligator habitat.[50] A somewhat amazing contrast exists between Langley's numbers for Louisiana and Florida, where population numbers of alligators are similar. Whereas the number of bites is high in Florida, the number is quite low in Louisiana. Several factors undoubtedly contribute to this contrast, the most important of which may be that human populations in Louisiana are not so integrated with alligator habitat. In addition, alligators in Louisiana appear to be more wary of humans than those in Florida.

48. McIlhenny, *Alligator's Life History.*
49. Ricky L. Langley, "Alligator Attacks on Humans in the United States," *Wilderness and Environmental Medicine* 16 (2005).
50. "Alligator Bites on People in Florida," Florida Fish and Wildlife Conservation Commission, updated December 2022, https://myfwc.com/media/1716/alligator-gatorbites.pdf.

Table 2. Alligator Bites on People in Florida

Decade	Major	Fatal	Minor	Total
2021–2022	16	(0)	4	20
2011–2020	74	(4)	28	102
2001–2010	76	(12)	30	106
1991–2000	56	(3)	35	91
1981–1990	42	(4)	43	85
1971–1980	44	(3)	1	45
1948–1968	4	(0)	0	4
Totals	312	(26)	141	453

Based on data from Florida Fish and Wildlife Conservation Commission, Updated December 2022

Wherever human activities occur in close proximity to alligator habitat, the potential for alligator attacks increases. Young children are especially vulnerable to attacks because of their smaller size and should be closely supervised or kept away from waters where alligators are common. Small animals such as dogs are even more vulnerable because they resemble other species that are natural prey of alligators, such as raccoons. Dogs walked along the edge of alligator habitat—or swimming in waters where alligators occur—are often attacked, resulting in danger to their owners who may try to rescue them.

Residential parks, university campuses, and similar institutions within the natural range of alligators often have substantial populations of alligators if suitable waterways are present. Observing alligators and other wildlife in such natural or semi-natural environment can be very rewarding, but care must be taken to be aware of one's surroundings, and especially of children and pets. Never feed alligators and be especially vigilant in locations where other people may have fed them. Alligators can become truly dangerous when they are fed by people and learn to associate humans with food. In locations where people are in close contact with alligators, such as in tourist attractions or modern housing developments near alligator habitat, this can become a serious problem, especially to small children and pets. People may be tempted to feed alligators as a means of drawing them closer for observation.

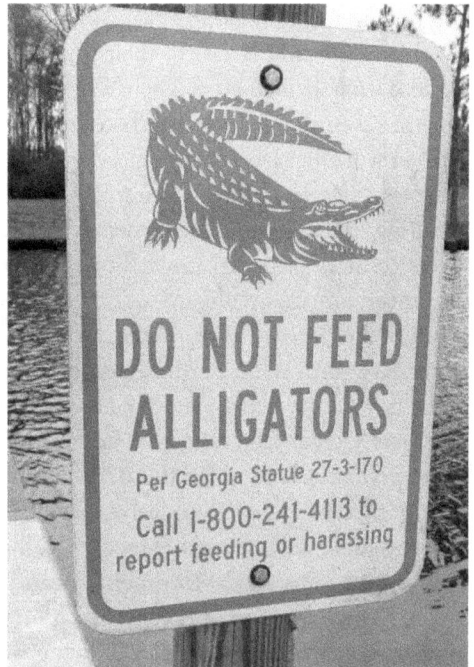

Alligator warning signs at (*left*) Myakka River State Park, Florida, January 20, 2018, and (*right*) Okefenokee National Wildlife Refuge, Georgia, February 10, 2019

When I was a student at the University of Florida in the 1960s, with its caged "Gator" mascot, Albert, a popular diversion from academics was to feed Albert and other alligators that resided in sinkhole ponds and Lake Alice on the campus, a practice that is now illegal. The most popular treat for both student feeders and alligator consumers was marshmallows, but certainly not the most natural and nutritious food for a reptile. There are also alligators in Cypress Lake on the campus of the University of Louisiana at Lafayette that are commonly fed by students, even though this practice is discouraged by the university administration. The University of Miami recently reported American crocodiles in the waterways on campus. These occurrences demonstrate that virtually any body of water within the range of alligators may potentially provide habitat for these reptiles and attract people to observe them. Even more appropriate foods such as chicken parts or fish create problems when fed to wild alligators. Alligators fed anything tend to become less fearful of humans and therefore more aggressive and dangerous. They may then become "nuisance alligators," requiring removal and potentially death. Many injuries or deaths (to both people and their pets) have resulted from attacks by such aggressive alligators. Today, the feeding of wild alligators is legally banned in virtually all areas of their range.

Chapter Five

Threats to and Abuse of Alligators

Based on the available sources, Native Americans seem to have presented little threat to alligators. Their limited hunting of alligators for food, the protective, somewhat inaccessible habitat of the animal, and the ceremonial or religious practices of many Native groups inhibited excessive killing and significant population decline (and probably minimized the mortality of humans from alligator attacks). This balance seems to have continued for thousands of years, until Europeans invaded North America in greater and greater numbers. Eventually (by about 1700) these new settlers had caused significant negative impacts on many species of wildlife, including alligators (as well as on the Native American people through disease, war, and displacement).

Several factors made these new immigrants to North America a serious danger to alligators. The increased human population, their obvious fear of alligators, their use of deadly firearms, the development of settlements within and adjacent to alligator habitat, the destruction of those habitats, and the development of commercial markets for alligator products in the 1800s caused significant increases in the mortality of the species.

A Commercial Frenzy

Undoubtedly the greatest mortality to the species came with the recognition that alligators had significant economic value and the resulting development of commercial markets for tanned alligator hides, first recorded in 1800. John J. Audubon, writing in 1827, noted that thousands of alligators were killed to satisfy a "mania of having either shoes, boots, or saddle-seats, made of their hides."[1] This temporary "mania" ended when the quality of alligator leather was considered inferior to other types of leather. In addition to the hides, many alligators were killed to produce lubricating oil for greasing the machinery of steam engines and cotton mills.

1. Audubon, "Observations," 3.

Charles H. Stevenson, with the US Fish Commission, wrote that the year 1855 kicked off another period of considerable demand for alligator leather, but this fashion demand was short-lived and ended after only a few thousand hides were shipped from the Gulf states.[2] However, demand soon resumed during the Civil War (1860 to 1865), when alligator hides were used to produce boots for Confederate troops because more desirable leathers were in limited supply. According to Stevenson, there was another brief lull in hunting pressure until 1869 when "fickle fashion" again demanded alligator leather for fancy slippers, boots, traveling bags, belts, cardcases, music rolls, and other items.[3] Improved commercial tanning techniques in the late 1800s also resulted in alligator leather of higher quality, both in terms of durability and aesthetic appeal, and created greater demand.[4] In addition to shoes and boots, manufacturers reportedly used alligator leather for saddles, trunks, traveling bags, purses, pocketbooks, upholstery, and book covers. Tourist novelties such as live and stuffed juvenile alligators and teeth also became popular, and markets for alligator meat increased. An 1891 report to the US Fish Commission estimated that more than 2,500,000 alligators had been killed in Florida alone between 1800 and 1891.[5]

By 1902, alligator populations in Louisiana and Florida had been reduced by about 80 percent during the previous twenty years.[6] Numbers had greatly decreased in all the southern states, forcing commercial tanneries in the United States to import skins from Mexico and Central America. Although Stevenson did not identify specific species—referring to all of them as alligators—those from Mexico and Central America would have been other crocodilian species (spectacled caiman, American crocodile, and Morelet's crocodile), as American alligators are extremely rare south of the Rio Grande.[7] However, Stevenson did recognize differences in the skins from different locations and noted that the "skins of the alligators or caymans [sic] from Brazil, Venezuela, and other South American countries" were of little

2. C. H. Stevenson, *Utilization of the Skins of Aquatic Animals*, US Fish Commission Report for 1902 (US Government Printing Office, 1904).

3. Stevenson, *Utilization of the Skins*, 342.

4. Mary Nickum, et al., "Alligator (*Alligator mississippiensis*) Aquaculture in the United States," *Fisheries Science & Aquaculture* 26, no. 1 (2018).

5. H. M. Smith, *Notes on the Alligator Industry*, Bulletin of the US Fish Commission for 1891 (Washington, DC: US Government Printing Office, 1893), 343–45.

6. Stevenson, *Utilization of the Skins*, 344.

7. L. Sigler, et al., "Searching for the Northern and Southern Distribution Limits of Two Crocodilian Species: *Alligator mississippiensis* and *Crocodylus moreletii* in South Texas, US, and in Northern Tamaulipas, Mexico," *Crocodile Specialist Group Newsletter* 26, no. 3 (2007).

value and did not come on the market in the United States.[8] Of an average of 280,000 skins per year processed in the United States, 56 percent (156,800) came from Mexico and Central America, and thus were not *Alligator mississippiensis* at all. Of the 123,200 American alligator skins processed per year, 22 percent (27,104) came from Florida, 20 percent (24,640) from Louisiana, and 2 percent (2,464) from other states. Stevenson concluded in this 1904 report that within a few years their scarcity would make alligator skins too expensive for general commercial markets.[9]

Sport Shooting Without Limits

Stevenson also complained that thousands of alligators were slaughtered for "sport" with no beneficial use made of them and attributed the population decline prior to 1902 largely to this "wanton sport."[10] In fact, this had been a common occurrence since the arrival of European explorers in the sixteenth century. William Bartram, while traveling through what is now Alachua County, Florida, was awakened at midnight to observe his traveling companions attacking (described as "a rare piece of sport") a twelve-foot alligator that had wandered into their campsite, lured by the scent of fish.[11] Some threw firebrands at the alligator's head while others "formed javelins of saplings, pointed and hardened with fire," which "they thrust down his throat into his bowels." After eventually growing tired of this "diversion and pleasure of exercising their various inventions of torture," they put him out of his misery

Timucua Indians of north Florida killing alligators by Theodor de Bry, 1591. *Source*: Florida Memory, State Library and Archives of Florida.

8. Stevenson, *Utilization of the Skins*, 343.
9. Stevenson, *Utilization of the Skins*.
10. Stevenson, *Utilization of the Skins*, 344.
11. Bartram, *Travels Through North & South Carolina*, 210.

with a rifle ball. Bartram's description of this attack with sharpened saplings is reminiscent of a 1591 engraving by Theodor de Bry of the Timucua of north Florida killing very large alligators.[12]

Such practices continued into the nineteenth and twentieth centuries. Biologist Kent Vliet provides a vivid description of this perverted "sport" and persecution of alligators in the early 1900s, including popular magazine stories of "alligator shooting" from steamboats along rivers, with no attempts to recover carcasses, which were left to rot.[13] Fifty years earlier, a number of Civil War soldiers from New York, Indiana, Kentucky, and Vermont recorded witnessing alligator shootings during the war. Kelby Ouchley suggests that such accounts were more commonly written by Union soldiers because they were less familiar with alligators than Southerners.[14] However, Confederate Major Ben McCulloch Hord from Tennessee published an account of alligator shooting at the close of the war, describing the event as affording "fine sport to a good marksman with a rifle."[15] Although participants fired "shot after shot," hitting many alligators and resulting in "violent lashing of the water," apparently the group recovered few skins (and likely caused the eventual deaths of most of the alligators hit).[16] As a result of this waste of impressive wildlife (and a valuable resource), Major Hord was able to go home to Tennessee with a twelve-foot trophy alligator skin.[17]

A Population in Decline

By the first half of the twentieth century, it seemed that the American alligator might go the way of other endangered species, such as the passenger pigeon, as their population numbers continued to decline. A keystone species in swamps and marshes, the alligator is important to the ecological balance of such environments. Where their numbers are reduced, or eliminated, the environment suffers. The alligator is a top predator and is important for controlling the numbers of its prey species. There is also an important aesthetic value of

---

12. Alexander, *Discovering the New World,* 43; engravings also available at State Archives of Florida, Florida Memory, https://www.floridamemory.com.

13. Vliet, *Alligators,* 36.

14. Ouchley, *American Alligator.*

15. Ben McCulloch Hord, "Alligator Shooting in Louisiana," *Trotwood's Monthly* 1–3 (1905): 250.

16. Hord, "Alligator Shooting in Louisiana," 250.

17. Unfortunately, such attitudes of misuse and waste still exist among some people today and occasionally result in dead alligators being seen floating in waterways adjacent to roadways, apparent victims of such ignorance and abuse.

living alligators: people enjoy seeing these iconic and impressive living relatives of the dinosaurs. Their loss would be a significant environmental tragedy.

Alligators, in their relatively secure river, swamp, and marsh habitats, initially suffered only limited negative impacts of hunting for meat and hides. But with increased habitat destruction, such as draining wetlands for farms and cities, suitable refuge habitat for alligators became more limited. Extensive threats to alligator populations began in the 1800s as humans began taking over their habitats, as well as increasing hunting pressure for hides. A bulletin for the US Department of Agriculture in 1929 noted that "continual drainage and encroachment of agricultural interests on the natural habitat of alligators, coupled with the wanton manner in which they have been hunted . . . caused a decrease in their numbers in many parts of their former range."[18] The report suggested that they had "retreated to the more unfrequented parts of large marshes and to inaccessible areas of extensive swamps." The ideal habitat for alligators became restricted to the Okefenokee Swamp, the Everglades, the cypress swamps of Florida, the marshes and swamps around Mobile Bay, the coastal areas of Mississippi, and the bayous and cypress swamps of Louisiana. According to the report's author, Remington Kellogg, if it were not for such areas, the collection of eggs, sale of young, and "reckless destruction of adults for belly skins" would eventually result in their extermination.

Stanley Arthur, director of the Wildlife Division of the Louisiana Department of Conservation, issued a more drastic warning: "It seems at this writing that this giant and characteristic reptile is doomed to disappear from our fauna—and within the next few years."[19] He further stated that at the rate of killing in 1927, the alligator "is doomed to certain extinction—and soon."

In the early 1900s, there were few restrictions on the killing of alligators, and only Florida and Louisiana gave them any type of legal protection or implemented license requirements for hunters.[20] Excessive killing and illegal poaching of alligators continued to be a serious problem. Stevenson estimated that the populations in Louisiana and Florida were less than 20 percent of what they were in 1880.[21] During the mid-1800s, a single hunter could kill thirty to forty alligators per night.[22] By about 1900, the success

18. Remington Kellogg, *The Habits and Economic Importance of Alligators*, US Department of Agriculture Technical Bulletin 147 (US Government Printing Office, 1929), 3. All subsequent quotations in this paragraph are taken from this report.
19. Arthur, "The Alligator," 165–66. The subsequent quotation is also taken from this source.
20. Kellogg, *The Habits and Economic Importance*.
21. Stevenson, *Utilization of the Skins*, 344.
22. Robert H. Chabreck, "The American Alligator—Past, Present and Future," *Proceedings of the Annual Conference of Southeastern Association of Game and Fish Commission* 21 (1967).

rate had declined to about ten to twelve per hunter per night, and by the late 1950s to about two to three per hunter per night, demonstrating the marked decrease in the species' numbers. Continued mortality from hunting resulted in further decline in the Louisiana population—90 percent by the 1940s to 1950s—and reached a record low in 1960.[23] Ross Allen noted that in Marion County, Florida, the location of his pioneering Reptile Institute, alligators were still abundant between 1931 and 1936, and hide hunters could take as many as twenty to fifty in a single night. By 1939–40, though, hunters had difficulty in finding ten to fifteen, and by 1943, even one or two.[24]

Similar or even greater declines had occurred in other parts of the alligator's range by the mid-twentieth century. By this time, American alligators were "practically non-existent" throughout much of its natural range except on wildlife refuges and other protected lands.[25] In describing the alligator population reduction in Florida, Georgia, and South Carolina, Allen and his associate Wilfred Neill used records of leading hide dealers in Florida to illustrate this decline, from a high of 190,000 hides processed in 1929 to a low of 6,800 in 1943 (Table 3).[26] They also pointed out that prices tended to rise as alligators became less common, resulting in greater hunting intensity as prices rose. Thus, greater mortality of already severely impacted populations caused further reduction in alligator numbers.

Although Allen and Neill noted that alligator populations in Florida, Georgia, and South Carolina showed marked recovery after initial legal protections were implemented in the mid-1940s, the general trend of population decline continued throughout most of the alligator range, reaching record low levels in the mid-1950s and early 1960s.[27] In fact, one reference book from 1957 noted that alligators had been "so badly persecuted that it now thrives only in the wilder parts of its original home," and that some states began to pass laws "to protect this valuable animal and save it from extinction."[28]

23. Chabreck, "The American Alligator."
24. E. R. Allen and W. T. Neill, "Increasing Abundance of the Alligator in the Eastern Portion of Its Range," *Herpetologica* 5, no. 6 (December 1949).
25. Chabreck, "The American Alligator," 556.
26. Allen and Neill, "Increasing Abundance of the Alligator," 109.
27. Ted Joanen, "Population Status and Distribution of Alligators in the Southeastern United States," unpublished report presented at Southeastern Regional Endangered Species Workshop, Tallahassee, FL, September, 1974.
28. Clifford H. Pope, *Reptiles Round the World* (Knopf, 1957), 57, 132.

Table 3. Alligator Hide Processing Records
for Leading Florida Hide Dealers

Year	Number of Hides Processed	Price Paid Per Best Hides
1929	190,000	$1.50
1930	188,000	$2.50
1931	150,000	$2.75
1932	145,000	$2.75
1933	130,000	$2.75
1934	120,000	$3.00
1935*	162,000	$3.00
1936	150,000	$3.00
1937	130,000	$4.00
1938	110,000	$4.00
1939	80,000	$5.25
1940	75,000	$7.00
1941	60,000	$8.75
1942	18,000	$15.75
1943	6,800	$19.25
1944**	7,000	$21.00
1945	12,000	$22.75
1946	10,000	$15.75
1947	25,000	$13.30

*Florida hide hunters began to move into Georgia in 1935.
**Laws passed in Florida in 1944 protected all alligators during the breeding season and banned the taking of those less than 4 feet (1.2 meters) long throughout the year.

Source: E. R. Allen and W. T. Neill, "Increasing Abundance of the Alligator in the Eastern Portion of Its Range," *Herpetologica* 5, no. 6 (December 1949).

Chapter Six

Initial Protection and Conservation of Alligators

Geneeral concern for the future of wildlife in North America began to grow in the early 1900s, resulting in the establishment of numerous wildlife refuges and sanctuaries in which all wildlife would be protected. One of the most significant federal actions that served to protect alligators and other wildlife was the creation of the US Fish and Wildlife Service (USFWS) in 1903 and its wildlife refuge program. Recognizing the need to provide greater protection for wildlife and their habitats, the USFWS established numerous national wildlife refuges in southeastern states in the first half of the twentieth century (Table 4), primarily for the protection of waterfowl and other migrant birds whose populations had been decimated. Given the overlap between the habitats of these waterfowl and the habitat of the American alligator, most of these refuges provided significant protection for both.

Louisiana

In Louisiana, alligator authority Edward A. McIlhenny was a leader in establishing several wildlife refuges.[1] McIlhenny was born in 1872 at Avery Island, a salt-dome island in the Louisiana coastal marshes and home of the McIlhenny Company, producer of Tabasco sauce. McIlhenny's father, Edmund McIlhenny, had founded the company and invented the now-ubiquitous pepper sauce. But living in the Louisiana marshes his entire life, as well as an enthusiastic and careful observer of nature, Edward McIlhenny became a respected authority on the wildlife of these wetland habitats, including alligators. An avid duck hunter and ornithologist, he also became one of the most devoted saviors of Gulf Coast wildlife.

---

1. K. M. Wicker, et al., *Rockefeller State Wildlife Refuge and Game Preserve: Evaluation of Wetland Management Techniques*, Coastal Management Section, Louisiana Department of Natural Resources, Baton Rouge, Louisiana, 1983.

**Table 4. National Wildlife Refuges Established
within the American Alligator Range Prior to 1970**

Year Established	State	Refuge Name
1920	Florida	Caloosahatchee
1927	Georgia-South Carolina	Savannah
1929	Florida	Cedar Keys
1930	Georgia	Wolf Island
1931	Florida	St. Marks
1932	South Carolina	Cape Romain
1932	North Carolina	Swanquarter
1934	North Carolina	Muttamuskeat
1935	Arkansas	White River
1936	Mississippi	Yazoo
1937	Florida-Georgia	Okefenokee
1937	Louisiana	Lacassine
1937	Louisiana	Sabine
1937	Texas	Aransas
1938	Alabama	Wheeler
1938	South Carolina	Tybee
1940	Georgia	Blackbeard Island
1940	Mississippi	Noxubee
1941	South Carolina	Santee
1943	Florida	Chassahowitzka
1945	Florida	Sanibel Island (Darling)
1946	Texas	Laguna Atascosa
1951	Florida	Loxahatchee
1956	North Carolina	Pocosin Lakes (Pungo)
1958	Louisiana	Catahoula
1962	Florida	Merritt Island
1962	Georgia	Harris Neck
1963	Texas	Anahuac
1964	Alabama	Choctaw
1964	Alabama-Georgia	Eufaula
1964	Florida	Lake Woodruff

EDWARD A. McILHENNY

Edward A. McIlhenny, ca. 1925.
Photo compliments of the E. A. McIlhenny Collection, Avery Island, LA.

Avery Island and surrounding marshes were McIlhenny's first conservation project. He established "Bird City" in 1895, a private bird sanctuary primarily to protect endangered snowy egrets, whose plume-like feathers were valued to adorn ladies' hats. Beginning with eight young egrets raised in an aviary at Avery Island, the resulting Bird City rookery supported an estimated 100,000 nesting birds by 1911. President Theodore Roosevelt, one of the

(*Above*) Edward A. McIlhenny and alligator nest, ca. 1925; (*below*) Edward A. McIlhenny
examining a large alligator in its study pen, Jungle Gardens, Avery Island, LA, ca. 1935.
Photos compliments of the E. A. McIlhenny Collection, Avery Island, LA.

most dedicated conservation presidents, referred to this private reserve on Avery Island as the "most noteworthy reserve in the country."[2] By 1935, the refuge had been expanded to 170 acres (68.8 hectares) and renamed Jungle Gardens, now recognized as a major tourist attraction.

Continuing his dedication to wildlife protection and conservation (which lasted throughout his life), Edward McIlhenny was at least partly responsible for over 175,000 acres (70,800 hectares) of wetlands preservation, providing protected habitat for alligators, waterfowl, and other wildlife. These included the Louisiana State Wildlife Preserve on Vermilion Bay in 1911, 13,000 acres (5,000 hectares) donated to the state by McIlhenny and C. W. Ward; Marsh Island Wildlife Refuge in 1920, 76,600 acres (31,000 hectares) donated by Mrs. Margaret Sage, at the urging of McIlhenny; and the Rockefeller Wildlife Refuge also in 1920, 86,000 acres (35,000 hectares) donated by McIlhenny in cooperation with the Rockefeller Foundation. These refuges were instrumental in reducing the slaughter of alligators, but unfortunately illegal hunting and poaching continued because restrictions were difficult to enforce in the remote habitats of the alligators.[3]

The state of Louisiana added additional protected habitat through its numerous wildlife management areas, and various nongovernmental conservation groups such as the National Audubon Society added more refuges as well. Their first refuge was created in Louisiana in 1924 on property donated to create the Paul J. Rainey Wildlife Sanctuary.[4]

Florida

The state of Florida also became a leader in protecting alligator habitats as refuges. The first of these, established in 1917, was designated the State Alligator Reservation on Tomoka Creek and River in Volusia County near Daytona Beach).[5] The law that created the refuge made it unlawful to "capture, injure or kill any alligators in the waters of, or along the shores of the Tomoka Creek and River" and that these areas were "declared a reservation for the protection of alligators."[6] Another version of this legislation was passed in 1931, making it unlawful "to shoot, kill, trap or otherwise molest alligators in Tomoka River" (See Table 5). Although this reservation was intended to

---

2. Theodore Roosevelt, *A Book-Lover's Holiday in the Open* (Charles Scribner's Sons, 1916), 42.
3. McIlhenny, *Alligator's Life History.*
4. "Paul J. Rainey Wildlife Sanctuary," Audubon Delta, accessed November 21, 2024, https://la.audubon.org/conversation/paul-j-rainey-wildlife-sanctuary.
5. Kellogg, *The Habits and Economic Importance*; Laws of Fla., ch. 7610, no. 352 (1917).
6. Laws of Florida, vol. 2, Chapter 7610—No. 352 (1917), 1850.

be a permanent refuge for the protection of alligators,[7] it unfortunately no longer exists, a victim of Florida's rapid population growth and commercial development. However, alligators in the area have been shielded since 1944 by statewide legal protection, and by federal protection since 1966. In addition, significant natural environments around Volusia County have been protected as part of Tomoka State Park, Addison Blockhouse Historical State Park, Tomoka Marsh Aquatic Preserve, Spruce Creek Preserve, and Spruce Creek Park. Though some habitat for alligators still exists along the Tomoka River, it has been largely replaced by the urban areas of Volusia County, with their residential developments, shopping centers, and golf courses.

Additional legislative actions implemented during the period between 1927 and 1939 indicate that Florida officials had begun to recognize the danger of unregulated killing of alligators but addressed the problem piece-meal with a confusing set of laws at the county level rather than statewide, or even regionally. According to Ross Allen, there was violent opposition to legal protection of alligators in Florida in 1935.[8] Despite this opposition, the legis-lature passed numerous state laws to prohibit the killing of alligators, mostly directed at specific counties, rivers, or lakes (Table 5), and which probably generated more confusion than protection for alligators. The first in 1927 made it "unlawful to capture, kill, catch, maim, injure, or shoot at or de-stroy alligators or alligator nests" in specified waters and adjacent marshes of Marion and Lake Counties (including Lake Dora, Dora Canal, Dead River, Lake Eustis, Haines Creek, Lake Griffin, Silver Springs Run, and Ocklawaha River).[9] A law passed in 1931 made it unlawful "to shoot, kill, trap or other-wise molest alligators in Spruce Creek, Volusia County," which is just south of Daytona Beach at Port Orange and New Smyrna Beach.[10] In 1933 and 1935, similar laws designated Silver River (also known as Silver Springs Run) and Ocklawaha River as protected. Elsewhere in Marion County in 1939, the capture of alligators for sale (plus fish, turtles, and frogs) was banned in Lake Weir and Little Lake Weir.

In 1937, the laws began to focus on specific Florida counties, begin-ning with Martin County, which prohibited "the Capture, Injury or Killing of Alligators, the Sale, Transporting and Transporting for Sale of Alligators, Alligator Skins, Alligator Teeth or Alligator Eggs."[11] This law also authorized the Florida Game and Freshwater Fish Commission to grant permits for

---

7. Kellogg, *The Habits and Economic Importance.*
8. Allen and Neill, "Increasing Abundance of the Alligator."
9. Laws of Fla., ch. 13068, no. 1263 (1927).
10. Laws of Fla., ch. 15489, no. 851 (1931).
11. Laws of Fla., ch. 18682, no. 976 (1937).

Tomoka River at (*above*) State Highway 40 and at (*below*) River Bend Nature Park, Ormond Beach, Florida, April 9, 2021

public or private zoos or parks in Martin County to exhibit captive alligators or for legitimate organizations exemptions for scientific purposes. Virtually identical wording was used in 1939 in laws to protect alligators in the counties of Broward, Charlotte, Dade, Indian River, and Orange, and more abbreviated language for Palm Beach and St. Lucie Counties (Table 5).

Then, in typically convoluted, political fashion, several additional laws were drafted that based protections upon county human population numbers according to the most recent census (1930 or 1935).[12] In 1937, this policy "Prohibit[ed] the Capture, Injury or Killing of Alligators, the Sale, Transporting and Transporting for Sale of Alligators, Alligator Skins, Alligator Teeth or Alligator Eggs" in "several" counties with human populations ranging from 7,150 to 7,200 (i.e., only Baker County).[13]

In 1939, the same prohibitions were applied to counties with populations of 12,960 to 13,000 (i.e., only Levy County, but it was repealed in 1941) and to counties with populations of 8,352 to 8, 400 (i.e., only Calhoun County). The legislature enacted similar prohibitions against selling alligators or alligator eggs in 1939 in counties with populations between 16,000 and 16,500 (i.e., Lee County) but also allowed the killing or capture of alligators longer than 30 inches (76 centimeters) between November 20 and February 20. Interestingly, these laws were ultimately applied using the 1935 Florida census, rather than the 1930 federal census.

In 1944, the Florida legislature passed laws that protected all alligators in the state during the breeding season and banned the taking of those less than 4 feet long (1.2 meters) throughout the year.[14] A law passed in 1950 protected alligators throughout the state, but in 1952, hunting was permitted for alligators 8 feet long (2.4 meters) and larger.[15] For comparison, Louisiana gave alligators partial protection in 1960, with a minimum size limit of 5 feet (1.5 meters). In 1962, both Florida and Louisiana established a total ban on the hunting of alligators due to continued declining numbers.[16]

12. There was some confusion at the time as to whether the 1930 federal census or the 1935 state census was used. In 1937, the "last Federal census" was specified (1930), but no county matches the designated population range. Baker County matches if the 1935 state census counts are used. In 1939, the state census of 1935 is specified. I have assumed that the 1935 census was used for all of them.

13. Laws of Fla., ch. 17709, no.901 (1937).

14. Allen and Neill, "Increasing Abundance of the Alligator."

15. Allan R. Woodward, "History of American Alligator Regulations in the U.S.A," Florida Fish and Wildlife Conservation Commission (June 12, 2007), https://myfwc.com/media/1742/alligator-regs-history.pdf.

16. Woodward, "History of American Alligator Regulations."

Table 5. Laws of Florida Addressing Alligator Protection

Date	Chapter (Number): Pages	Locations
1917	Ch. 7610 (No. 352): 452-453	Tomoka Creek and River in Volusia County
1927	Ch. 13068 (No. 1263): 2668-69	Several waters in Marion and Lake Counties
1931	Ch.15489 (No.851): 1615	Spruce Creek in Volusia County
1931	Ch.15550 (No. 912): 1850	Tomoka River in Volusia County
1933	Ch.16044 (No. 187): 382	Silver River and Ocklawaha River in Marion County
1935	Ch. 17031 (No. 260): 593	Silver River and Ocklawaha River in Marion County
1937	Ch. 17709 (No. 901): 26-28	The "several Counties" with populations ranging from 7,150 to 7,200 (= only Baker County based upon 1935 census)
1937	Ch 18682 (No. 976): 987-988	Martin County specified
1939	Ch.19266 (No. 271): 506	All Counties with population from 12,960 to 13,000 (= only Levy County based upon 1935 census) (Repealed in 1941 – Ch. 20359)
1939	Ch. 19581 (No. 586): 1433-34	All Counties with population from 16,000 to 16,500 (= only Lee County based upon 1935 census)
1939	Ch. 19620 (No. 625): 1496-97	All Counties with population from 8,352 to 8,400 (= only Calhoun County based upon 1935 census)
1939	Ch. 19706 (No. 711): 109-110	Broward County specified
1939	Ch. 19724 (No. 729): 218-219	Charlotte County specified
1939	Ch. 19765 (No. 770): 308-309	Dade County specified
1939	Ch. 19896 (No. 901): 744	Indian River County specified
1939	Ch. 19966 (No. 971): 1024-25	Lake Weir and Little Lake Weir in Marion County
1939	Ch. 20014 (No. 1019): 1239	Orange County specified
1939	Ch. 20037 (No. 1042): 1279	Palm Beach County specified
1939	Ch. 20112 (No. 1117): 1501	St. Lucie County specified

Source: Florida Legislature, 1917–1939. Laws of Florida, Chapters 7610–20112.

Virtually any freshwater body in Florida and coastal Louisiana could be expected to provide habitat for alligators, so the numerous state parks, state forests, and wildlife management areas also protected habitat for alligators. Of course, the creation of Everglades National Park in 1947 preserved more than 1.5 million acres (607,000 hectares) of habitat for alligators. Additional preserves, state forests, and wildlife management areas surrounding the national park, and the adjacent Big Cypress National Preserve, with over 700,000 acres (283,280 hectares) established in 1974, have greatly expanded protected habitat. The estimated number of alligators in all of Florida today is about 1.3 million.[17]

Other State Policies

Wildlife officials in other states also recognized the seriousness of the major decrease in alligator numbers in the early 1900s and became concerned that alligator populations had declined to dangerously low levels in much of their natural range. Most of these states implemented a variety of regulations by the early 1960s that provided some protection to alligators. The state of Alabama was the first to give them complete protection, passing such a law in 1938. In Georgia, alligators were completely protected by law in the 1940s, protection which continued until 1987.[18] South Carolina banned night shooting of wildlife in 1955, which indirectly protected alligators.[19] Then, beginning in 1962, alligator trappers in the state were required to have licenses and tags. The South Carolina alligator hunting season was closed in 1964 and was not reopened until 1995. Arkansas provided legal protection for alligators in 1961.[20] There was no regulatory protection of alligators in Mississippi prior to 1967, in Texas prior to 1969, nor in North Carolina prior to 1973.[21]

Even in states where significant legal protections for alligators were implemented, the laws were difficult to enforce, resulting in poor compliance to

17. "A Guide to Living with Alligators," Florida Fish and Wildlife Conservation Commission, April 2024. https://myfwc.com/media/dmwbfdw4/alligator-brochure.pdf.
18. Allen and Neill, "Increasing Abundance of the Alligator," 110.
19. Walter E. Rhodes, "Conservation of a Dinosaur in Modern Times: South Carolina's Alligator Management Program," *Eighth Eastern Wildlife Damage Management Conference* 8 (1997), 126.
20. Mark Barbee, *Alligator Hunt Orientation and Training Manual*, Arkansas Game and Fish Commission (2020), 3.
21. Personal communication with Ricky Flynt, Mississippi Alligator Program Coordinator, October 2020; "American Alligator," Texas Parks and Wildlife, accessed February 24, 2025. https://tpwd.texas.gov/huntwild/wild/species/alligator/index.phtml, 1; *North Carolina Alligator Management Plan*, North Carolina Wildlife Resources Commission (2017), 7.

these regulations in some areas and continued illegal hunting and poaching. However, in 1969, Congress amended a federal law known as the Lacey Act (passed in 1900), which prohibited the transport of illegally obtained game birds and mammals across state lines, to include reptiles. This action contributed significantly to the recovery of the American alligator.[22] By prohibiting the transport of illegally obtained alligator skins from one state to another, this law, with harsh penalties for violations, effectively closed skin markets in New York, which had depended upon poaching to supply their raw materials. These limited protective measures, along with other management projects, stimulated significant recovery of alligator populations, and numbers began to rapidly increase.

In the 1970s, various states initiated intensive research on alligator populations and began, in recognition of significant population increases, to implement management plans allowing limited commercial "harvesting" of alligators while also protecting them from excessive killing. Credit must be given to the team of Ted Joanen and Larry McNease and other Louisiana Department of Wildlife and Fisheries (LDWF) biologists who, in the 1970s and 1980s, led the way in scientific studies to develop sustainable harvest programs to ensure the recovery of alligator populations.[23] Management plans developed in Louisiana have been the model for other states to follow in helping to protect alligators.

Based on the assessment of the first legal alligator harvest, held in Cameron and Vermilion Parishes, Louisiana, in 1972, one report noted that the local population increased the following year by 30%, despite the harvesting of more than 1,100 alligators.[24] The report's author, Bob Dennie, confirmed several prior assumptions based upon this experimental harvest: alligators are a renewable resource; harvesting alligators can provide significant annual revenue to the local economy; such revenue provides incentive to protect local alligator populations; and local marshland (i.e., alligator habitat) is much more valuable for renewable alligator harvests than drained for agricultural production (thus providing increased incentive for environmental protection).

22. Rhodes, "Conservation of a Dinosaur."
23. Ted Joanen and Larry McNease, "The Management of Alligators in Louisiana, USA," in *Wildlife Management: Crocodiles and Alligators*, eds. J. W. W. Grahame, et al. (Surrey Beatty & Sons, 1987).
24. Bob Dennie, "Gatortime—Part Two," *Louisiana Conservationist* 25 (September-October 1973).

Table 6. Alligator Population Estimates, 1973

State	No. of Alligators	Habitat Size in Square Miles (sq. km)	Alligators per Square Mile
Florida	407,585	26,852.00 (69,546)	15.18
Louisiana	200,682	7,992.14 (20,700)	25.11
South Carolina	48,700	3,660.00 (9,479)	13.31
Georgia	29,954	2,985.27 (7,732)	10.03
Texas	26,784	2,029.00 (5,247)	13.2
Alabama	12,715	272.88 (707)	46.6
Mississippi	4,740	288.79 (748)	16.4
Arkansas	1,900	11.72 (30)	162.1
North Carolina	1,314	1,014.10 (2,626)	1.3
Oklahoma	10	156.25 (405)	0.06
Totals	734,384	45,262.15 (117,228)	16.23

Table 7. Estimated Alligator Population Trends by County, 1973

State	Number (%) of Counties Increasing	Number (%) of Counties Stable	Number (%) of Counties Decreasing
Florida*	-	-	-
Louisiana	40 (63%)	22 (35%)	1 (1.6%)
South Carolina	10 (36%)	18 (64%)	0
Georgia**	48 (56%)	32 (38%)	5 (6%)
Texas	35 (56%)	22 (35%)	5 (8%)
Alabama	15 (54%)	11 (39%)	2 (7%)
Mississippi	13 (24%)	38 (69%)	4 (7%)
Arkansas	3 (100%)	0	0
North Carolina	4 (19%)	9 (43%)	8 (38%)
Oklahoma	0	0	1 (100%)
Totals	168 (48%)	152 (44%)	26 (7%)

*Information not reported for Florida.
**Ten counties undetermined in Georgia.

Source: Ted Joanen, "Population Status and Distribution of Alligators in the Southeastern United States," unpublished report presented at Southeastern Regional Endangered Species Workshop, Tallahassee, FL, September 1974.

In 1973, Joanen led a team of wildlife biologists from the Southeast to estimate population numbers in each state (Table 6).[25] He noted that alligator populations had dramatically increased by the 1970s compared to the all-time low levels in the mid-1950s and early 1960s, with an estimated population total of 734,384 alligators living on 45,000 square miles (117,000 square kilometers) of available habitat. Florida and Louisiana continued to support the largest populations, but all states with significant alligator habitat had either stable or increasing populations based upon counties surveyed (Table 7).

In Louisiana, Joanen and McNease demonstrated a clear pattern of increasing population in coastal marshes annually between 1970 and 1983.[26] They counted the number of nesting females from aerial nest surveys and then multiplied the results by twenty to estimate the total population (assuming nesting females made up 5 percent of any given population).[27] However, later studies indicated that actual populations were much larger, possibly as much as four times the previous estimates.[28] Consequently, later LDWF reports presented data as number of nests counted.[29] However, the aerial nest survey method is still valid for determining population trends. The population increase witnessed during the 1970s and early 1980s was especially significant, as it coincided with the Louisiana initiative to allow experimental alligator hunts, resulting in a total of 100,712 harvested during the ten-year period from 1972 to 1983 (Table 8).[30] Annual nest surveys have continued by LDWF personnel to the present day, indicating that populations have continued to increase (Figure 2). Despite a period of apparent leveling of the population in the 1990s, the state witnessed another period of increase between 2012 and 2020. The nest count for 2019 (67,900) was the highest on record.

25. Joanen, "Population Status and Distribution of Alligators."
26. Joanen and McNease, "The Management of Alligators."
27. Larry McNease and Ted Joanen, "Distribution and Relative Abundance of the Alligator in Louisiana Coastal Marshes," *Proceedings of the Annual Conference of Southeastern Associations of Fish and Wildlife Agencies* 32 (1978); Robert H. Chabreck, "Methods of Determining the Size and Composition of Alligator Populations in Louisiana," *Proceedings of the Annual Conference of Southeastern Associations of Game and Fish Commissions* 20 (1966).
28. Dave Taylor, et al., "Female Alligator Reproduction and Associated Population Estimates," *Journal of Wildlife Management* 55 (April 1991).
29. *Louisiana's Alligator Management Program 2021–2022 Annual Report*, Louisiana Department of Wildlife and Fisheries (2022).
30. Joanen and McNease, "The Management of Alligators."

Table 8. Alligators Taken in Louisiana's
Experimental Hunting Seasons, 1972–1983

Year	No. of Tags Issued	No. of Alligators Taken	Value of Skins*
1972	1,961	1,350	$75,670
1973	3,243	2,921	$290,714
1975	4,645	4,420	$261,570
1976	4,767	4,389	$515,003
1977	5,760	5,474	$492,061
1979	17,516	16,300	$1,691,940
1980	19,134	17,692	$1,515,674
1981	15,534	14,870	$1,800,757
1982	18,188	17,142	$1,578,264
1983	17,130	16,154	$1,453,214
Total	107,878	100,712	$9,674,867

*Note: Skin values listed in Joanen and McNease are slightly different from the LDWF values.

Sources: Ted Joanen and Larry McNease, "The Management of Alligators in Louisiana, USA," in *Wildlife Management: Crocodiles and Alligators*, eds. J. W. W. Grahame, et al. (Surrey Beatty & Sons, 1987); *Louisiana's Alligator Management Program 2021– 2022 Annual Report*, Louisiana Department of Wildlife and Fisheries (2022).

The Endangered Species Act

The ultimate protection of American alligators came in 1973 with passage of the federal Endangered Species Act. Other, less comprehensive measures preceded this law, including the Endangered Species Preservation Act of 1966, which, in conjunction with the species list published in the Federal Register,[31] classified the American alligator (as well as all other listed species) as endangered (defined as species in danger of extinction within the foreseeable future throughout all or a significant portion of its range). However, it did not include regulations to protect the listed species except on federal lands, such as national wildlife refuges and national forests. It did provide the comprehensive definition of the terms

31. Native Fish and Wildlife Endangered Species, 32 Fed. Reg. 4001 (March 11, 1967).

Figure 2. Nest Count Estimates Based on Annual Aerial Surveys of Louisiana Coastal Marshes, 1970–2021

Nest Counts

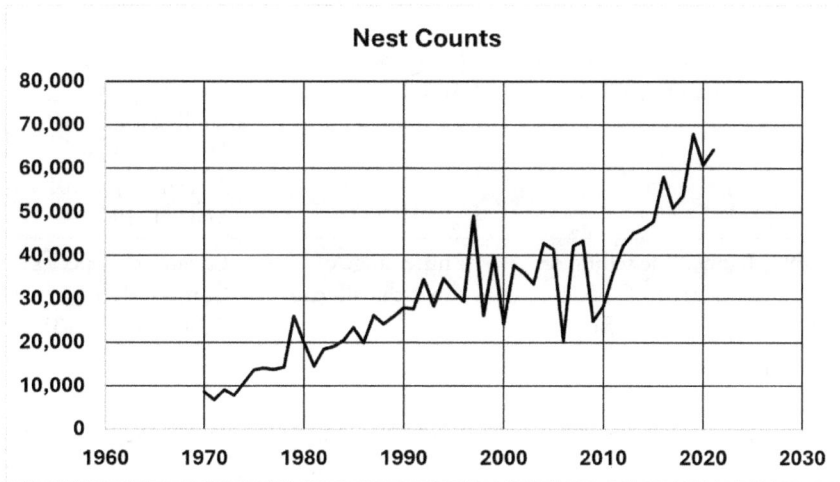

Source: Louisiana's Alligator Management Program 2021–2022 Annual Report, Louisiana Department of Wildlife and Fisheries (2022), 5.

take, taking, or *taken* (to pursue, hunt, shoot, capture, collect, kill, or attempt to pursue, hunt, shoot, capture, collect, kill) that have become part of subsequent endangered species regulations. The law also authorized the US Fish and Wildlife Service to purchase additional lands to preserve critical habitat for listed species. Subsequently named the Endangered Species Conservation Act, which was amended in 1969, it provided increased protection by banning the selling or importing of species recognized as facing worldwide extinction.

Then, in 1973, Congress passed the Endangered Species Act of 1973 with overwhelming enthusiasm, as well as the support of President Richard Nixon. This new law provided full protection to listed endangered species and also established the new classification of "threatened" (defined as species likely to become endangered within the foreseeable future throughout all or a significant portion of its range). The updated list included the American alligator, which made it illegal to kill or otherwise harm (or "take") alligators everywhere throughout the United States. In 1975, the American alligator gained additional international protection with approval of the Convention on International Trade in Endangered Species of Wild Fauna and Flora (CITES).[32]

32. Woodward, "History of American Alligator Regulations."

Table 9. Changes to Public Laws Regarding the American Alligator

Year	Action Taken on Alligators	Area Affected	References*
1967	Listed as threatened with extinction (endangered)	Entire range of American alligator	Endangered Species Preservation Act, Public Law 89-669, Oct. 15, 1966; 32 Fed. Reg. 4001 Mar. 11, 1967
1969	Banned the selling or importing of species recognized as facing worldwide extinction (endangered)	Entire range of American alligator	Endangered Species Conservation Act, Public Law 91-135, Dec. 5, 1969
1973	Listed as endangered	Entire range of American alligator	Endangered Species Act of 1973, Public Law 93-205, Dec. 28, 1973
1975	Reclassified from endangered to T/SA**	Cameron, Vermilion, and Calcasieu Parishes, Louisiana	40 Fed. Reg. 44412, Sept. 26, 1975
1977	Reclassified from endangered to threatened	All of Florida and coastal areas of Georgia, Louisiana, South Carolina, and Texas	42 FR 2071, Jan. 10, 1977
1979	Reclassified from endangered to T/SA	Nine additional parishes of Louisiana	44 Fed. Reg. 37130, June 25, 1979
1981	Reclassified from endangered to T/SA	Fifty-two additional parishes of Louisiana	46 Fed. Reg. 40664, Aug. 10, 1981
1983	Reclassified from endangered to T/SA	Entire range in Texas	48 Fed. Reg. 46332, Oct. 12, 1983
1985	Reclassified from threatened to T/SA	Entire range in Florida	50 Fed. Reg. 25672, June 20, 1985

| 1987 | Reclassified from endangered or threatened to T/SA in Georgia and South Carolina; and reclassified from endangered to T/SA in Alabama, Arkansas, Mississippi, North Carolina, and Oklahoma | Remainder of alligator range in states of Alabama, Arkansas, Georgia, Mississippi, North Carolina, Oklahoma, and South Carolina | 52 Fed. Reg. 21059, June 4, 1987 |

*References in US Federal Register (also see 86 Fed. Reg. 5112, Jan. 19, 2021 for additional changes)
** T/SA = "Threatened (Due to Similarity of Appearance)"

During the next few years, several additional amendments modified the Endangered Species Act (Table 9), allowing for some flexibility in the management of alligator populations. The downlisting of alligators to threatened in parts of its range in 1975 allowed states, led by Louisiana, to begin more extensive research and management programs that greatly enhanced our collective scientific knowledge of alligators. These included experimental harvesting with strict regulations for tagging and reporting requirements and quotas to ensure stable or increasing populations. Exemptions to control nuisance alligators were also added. Markets for skins, meat, and other products are now strictly monitored and controlled.

The Endangered Species Act has faced both praise and vilification as a powerful force for environmental protection. Although most citizens value nature and recognize the need to protect endangered species and their habitats, when protections threaten economic benefits, either real or imagined, then opposition to such protections can become significant. Consequently, the law has faced challenges and numerous amendments to modify and weaken it. Such amendments now allow exemptions to the law if protecting a species will result in significant negative economic impacts. New listings of species are frequently challenged by business interests. Recovery plans must be developed and monitored by the US Fish and Wildlife Service, and species that have recovered are delisted or downlisted. This has occurred with the American alligator, which was reclassified beginning in 1977 in several regions where it had become common and finally downlisted from endangered to threatened "due to similarity of appearance" throughout its range in 1987. The latter phrase was added to provide protection to other similar endangered species and (initially) to protect populations of American alligators in areas where the species was still considered endangered or threatened.

The American Alligator Today

Today, the estimated population of wild American alligators is 3 to 4 million non-hatchlings, of which approximately 26.5 percent (795,000 to 1,060,000) are adults.[33] Although now considered common and fully recovered, the American alligator remains categorized as threatened due to similarity of appearance chiefly for the benefit of other species of crocodilians, such as the American crocodile (which is still listed as threatened).

Some have argued that the American alligator was never in any real danger of extinction—even in the early 1970s—because estimated numbers were still in the hundreds of thousands, and in refuge areas they were abundant and protected. Even Ted Joanen emphasized that in 1974 alligator populations had "made dramatic increases when compared to the record low population levels of the mid-1950s and early 1960s."[34] He stressed that a simple, region-wide management plan was needed, including management of alligator populations as a renewable resource with sustained yield harvesting. He reasoned that if private landowners could benefit financially from alligators "harvested" on their lands, they would be inclined to protect alligator habitat, avoid over-harvesting, and help maintain a stable population. He feared that intense, unscientific conservation measures could deter proper state management programs. Thanks to the wisdom and expertise of Joanen and his associates in Louisiana, as well as his influence on wildlife agencies in other states, the American alligator would likely not have become extinct even without the support of the Endangered Species Act. However, this cannot be stated for certain, if one considers how extreme the decimation of alligator populations had been. In much of their range, even in parts of Louisiana, alligators had disappeared. The focus of the Act on alligators and other species that had suffered major decreases in population numbers drew national attention to the critical need for increased protection of such species and their habitats.

In their defense, the wildlife experts in Louisiana and other states had good data to support the idea that the alligator would never have become extinct, especially with the management programs that had already begun to increase their population size. But in other parts of their range, alligators had reached dangerously low levels that threatened their continued existence without proper protection.

33. Ruth Elsey, et al., *"Alligator mississippiensis,"* IUCN Red List of Threatened Species 2019, accessed November 25, 2024, http://dx.doi.org/10.2305/IUCN.UK.2019-2.RLTS.T46583A3009637.en.
34. Joanen, "Population Status and Distribution," 1.

There is no doubt that the Endangered Species Act, in combination with both state and federal management and conservation programs, has been a very effective vehicle to help protect many threatened and endangered species. If nothing else, the focus of the Endangered Species Act on alligators—and other species that had suffered major decreases in population numbers—drew national attention to the critical need for increased protection of such species and their habitats. No one in the early 1800s would have expected the abundant passenger pigeon to become extinct, and yet it was gone by 1914. This edible bird had significant economic value that should have ensured its survival but unfortunately did not. The same could have been the fate of the American bison that roamed the Great Plains in incredible numbers but was almost wiped out by excessive killing before protective measures, aided in some cases by its economic value, helped to ensure its survival and recovery. In the same way, the American alligator has survived because of the combined effects of endangered species protection, science-based management plans, and proper development of its economic potential—actions which were implemented before its demise had reached a point that protection was too late and recovery impossible. For both species, the dual incentives of ecological and economic value worked together to help ensure that their populations would recover. Of course, we should avoid valuing any species, including alligators, merely upon monetary factors, since their ecological and aesthetic value may be even more significant.

More recently, conservative commentators have attempted to devalue and discredit conservation efforts, in some cases referencing the American alligator specifically. Researcher and consultant Brian Seasholes expressed a more extreme opinion in opposition to the Endangered Species Act in 2013, labeling the federal recovery plans for alligators "the Great Gator Hoax."[35] Although based to a large extent on Ted Joanen's research and opinions, Mr. Seasholes's major goal was obviously to discredit the Endangered Species Act rather than to express concern for protecting alligators or other endangered and threatened species. His claim that the Endangered Species Act has caused more harm than good is not supported by clear evidence. The successful recovery of numerous iconic species including the American bison, bald eagle, brown pelican, and California condor, as well as the American alligator, provides strong confirmation that the Endangered Species Act has not caused harm to these conservation efforts. The listing of the American

35. Brian Seasholes, "The Great Gator Hoax," Green Watch, Capital Research Center (Washington, DC), February 8, 2013, https://capitalresearch.org/app/uploads/2013/02/GW1302-final-1301181.pdf.

alligator as an endangered species was not a "hoax" but rather a clear effort to provide needed protection for a species that had been severely depleted in most of its range. Seasholes's affiliation with conservative economic policy organizations, including the Competitive Enterprise Institute and Property and Environmental Research Center, demonstrates a clear bias in favor of economic policies, often at the expense of threatened and endangered species. Their policies also show a bias against other environmental protection issues in favor of monetary gain. To such an argument, Edward McIlhenny's assessment of alligator abuse (in 1935) bears repeating: "That the alligator has already been exterminated over a large portion of its former habitat is a fact, and one that civilization should not be proud of." Our civilization can take pride in the successes of the Endangered Species Act.

Chapter Seven

Controlled Harvests and Population Management

Though the Endangered Species Act provides significant protection for threatened animals across the country, it nonetheless also grants exemptions to states for research and experimental harvesting, which has allowed a much greater understanding of American alligator biology and factors controlling their population numbers. With the official recovery and downlisting of the alligator in 1987, states enacted additional management plans and enhanced those that had been developed during the previous decades of the alligator recovery process. Controlled harvesting seasons for wild alligators and captive rearing on alligator farms or ranches constitute two of the most significant changes in the way alligators have been managed. Both methods (discussed in this and the following chapter) are based upon the concept that alligators can be an economically beneficial and renewable resource if properly protected and managed, in contrast to the uncontrolled slaughter prior to the 1970s or the complete bans on killing alligators as an endangered species.

LDWF led the way with its 1972 Experimental Alligator Harvest Program. State law classified the alligator as a non-game, commercially valuable species, and the objective of the Experimental Harvest Program was to enhance alligator populations while also providing for the "harvest of surplus animals on a sustained yield basis."[1] In this context, "surplus" was used to designate the number of animals greater than that needed to maintain a stable, sustained yield harvest during subsequent years, initially set in 1972 at 20 percent of the experimental population. In 2021, the current annual harvest was reported to be approximately 3 percent of the Louisiana population.[2]

The first experimental harvest in 1972 was deemed a success: hunters claimed 1,350 alligators during the thirteen-day season in September with no apparent adverse effects on population levels (see Table 8 in previous chapter).

1. A. W. Palmisano, et al., "An Analysis of Louisiana's 1972 Experimental Alligator Harvest Program," *Proceedings of the Annual Conference of Southeastern Associations of Game and Fish Commissioners* 27 (1973): 184 (see also 191).
2. Ted Joanen, et al., "Evaluation of Effects of Harvest on Alligator Populations in Louisiana," *Journal of Wildlife Management* 85, no. 4 (2021): 1, 4.

In subsequent years, Louisiana increased the number of permits issued based on estimated population numbers. Through 1983, the number of alligators killed was less than the number of permits issued, so not all of the "surplus" alligators were being harvested. Hunters harvested a total of 100,712 alligators in that time, with a value of $9,821,277 for skins.[3] When the state legalized the sale of meat in 1979, additional economic value estimated at $125,000 per year was added, and the sale of parts such as teeth and skulls added even more. By 2022, more than 1.1 million wild alligators had been harvested in Louisiana, with an estimated value over $205 million for skins and over $134 million for meat, or $339 million total (Table 10).[4] Various novelty items, such as mounted heads, feet, and teeth, and fees for professional hunting guides hired to lead novice alligator hunters further increased the value of these harvests.

Of critical importance to the success of the alligator hunting season in Louisiana and other states has been the strict regulations and monitoring required by the programs. Annual population surveys establish harvest limits, hunts are restricted to the month of September to limit mortality of mature females (which tended to remain in interior marshes with limited access by hunters), hunters face strict licensing and tagging requirements, and the state conducts subsequent monitoring of commercial sales of skins, meat, and other parts.[5]

Florida, with its burgeoning populations of both alligators and humans, began its harvesting programs as a nuisance alligator control method in 1977, following the downlisting of alligators from endangered to threatened in the state.[6] In the 1970s, the Florida Fish and Wildlife Conservation Commission (Florida FWC) received four to five thousand complaints per year of nuisance alligators, as increasing numbers of human developments moved into close proximity to alligator habitat.[7] This nuisance alligator control program has become critical and continued up until the present with a recent average of more than seven thousand alligators taken each year.[8]

3. Ted Joanen, et al., "Louisiana's Alligator Management Program," *Proceedings of the Annual Conference of Southeastern Associations of Game and Fish Commissioners* 38 (1984): 206.

4. LDWF, 2022

5. Joanen and McNease, "The Management of Alligators."

6. Harry J. Dutton, et al., "Florida's Alligator Management Program: An Update, 1987 to 2001," in *Crocodiles: Proceedings of the 16th Working Meeting of the Crocodile Specialist Group* (International Union for the Conservation of Nature, 2002); Tommy Hines, "Alligator Harvest in Florida," in *Amphibians and Reptiles: Status and Conservation in Florida*, eds. W. E. Meshaka Jr. and K. J. Babbitt (Krieger, 2005).

7. Hines, "Alligator Harvest in Florida."

8. Alligator Management Team, "Table 1. Estimated Producer Value of Wild Alligator Harvests in Florida During 1977-2022," Florida Fish and Wildlife Conservation Commission (2023), https://myfwc.com/media/1775/alligator-wild-value.pdf.

Table 10. Wild Alligator Harvest Data, Louisiana and Florida

Louisiana Wild Harvest (1972–2022)				
Years	Number Taken	Skin Value	Meat Value	Total Value 1972–2022
1972–79	34,854	$3,326,958	$125,000	
1980–89	194,631	$58,924,857	$10,945,000	
1990–99	272,308	$57,192,295	$36,327,570	
2000–09	309,525	$59,683,590	$34,556,389	
2010–19	294,643	$42,490,847	$44,028,866	
2020–22	52,236	$3,437,490	$4,706,996	
Total	1,158,197	$205,056,037	$134,392,821	$339,448,858
Florida Wild Harvest (1977–2022)				
Years	Number Taken	Skin Value	Meat Value	Total Value 1977–2022
1977–79	4,085	$154,614	$14,400	
1980–89	39,711	$9,736,402	$5,989,720	
1990–99	94,842	$22,213,461	$10,726,631	
2000–09	163,013	$36,438,756	$15,034,588	
2010–19	176,158	$21,799,072	$30,400,538	
2020–22	54,087	$3,536,592	$12,316,494	
Total	531,896	$93,878,898	$74,482,371	$168,361,269

Sources: *Louisiana's Alligator Management Program 2021–2022 Annual Report*, Louisiana Department of Wildlife and Fisheries (2022); Alligator Management Team, "Table 1. Estimated Producer Value of Wild Alligator Harvests in Florida During 1977-2022," Florida Fish and Wildlife Conservation Commission (2023), https://myfwc.com/media/1775/alligator-wild-value.pdf.

Between 1981 and 1984, experimental harvest studies were conducted in several lakes in Florida and indicated that sustained harvests of 3–7 percent of a population could be implemented without seriously impacting the population.[9] Based on these findings, as well as expanded studies during the next few years, the state implemented a public hunt program that has continued up until the present, yielding an average harvest of more than seven thousand alligators

9. Hines, "Alligator Harvest in Florida."

per year.[10] This program has been described as primarily a "sport hunt," with restricted entry and a limit of two alligators for each permit holder.[11] In 1985, Florida added a third alligator harvest program on private land, with the objective of generating income for private landowners, an incentive to maintain a stable population of alligators on their property. This program has generated a harvest averaging more than two thousand alligators per year through 2022.

The wild alligator harvest in Florida from 1977 to 2022 yielded more than 531,000 hides, with an estimated value of almost $93,000,000 and meat value of more than $74,000,000 (for a total value estimated to be more than $168 million, see Table 10). Surprisingly, between 2013 and 2022, the value of alligator meat has exceeded the value of skins in Florida, apparently correlated with increased market demand for meat (possibly related to increased popularity of Cajun and Creole cuisine, whether actual or ascribed) and decreased demand for hides. Vaughn L. Glasgow, former curator at the Louisiana State Museum, described in detail the recent history of increased popularity of alligator as food.[12] The gourmet cook might also want to see the classic book by Ernest (Ernie) A. Liner, *The Culinary Herpetologist*, which includes 379 recipes for cooking alligator meat (or, rather, 378 for alligator and one for crocodile, with the footnote that alligator can be used).

Texas followed the lead of Louisiana and Florida, initiating a wild alligator hunting season in 1984. The Lone Star State has the distinction of being at the western limit of the American alligator range, with members of this species present only in the easternmost third of the state. There have been rare sightings of alligators on the Mexican side of the Rio Grande, but these are possibly the result of escaped captives.[13] For the purpose of regulating alligator hunts, the Texas Parks and Wildlife Department has designated twenty-two counties with the best alligator habitat as "core counties" (specifically, Angelina, Brazoria, Calhoun, Chambers, Galveston, Hardin, Jackson, Jasper, Jefferson, Liberty, Matagorda, Nacogdoches, Newton, Orange, Polk, Refugio, Sabine, San Augustine, San Jacinto, Trinity, Tyler, and Victoria).[14] Texas currently has two distinct alligator hunting seasons: one in the core counties in September, and the other from April through June in all other counties, where alligator habitat is marginal or nonexistent.

10.   Alligator Management Team, "Table 1."
11.   Hines, "Alligator Harvest in Florida."
12.   Glasgow, *A Social History of the American Alligator*, 92–105.
13.   Sigler, et al., "Searching for the Northern and Southern Distribution."
14.   "Alligators in Texas: Rules, Regulations and General Information, 2019–2020," Texas Parks and Wildlife Department, https://tpwd.texas.gov/publications/pwdpubs/media/pwd_bk_w7000_1011.pdf, accessed December 2, 2024.

In 1989, Georgia initiated a nuisance alligator program but only allowed licensed trappers to harvest alligators that presented a potential problem (or nuisance) to humans, averaging about 170 per year. Then, in 2003, Georgia began a regulated public hunting season to manage its flourishing alligator population. The total harvest between 2003 and 2018 was 3,383, or about two hundred alligators per year. Other states soon followed in enacting alligator hunts, including Mississippi (2005), Alabama (2006), Arkansas (2007), and South Carolina (2008). These hunts have generally yielded a harvest of several hundred alligators per year. These states had also enacted nuisance alligator programs as increasing populations generated more frequent conflicts with increasing human populations.

Virtually all state programs except for the nuisance programs and commercial programs on private lands in Louisiana and Florida could be regarded as sport or recreational hunts, providing additional economic benefits. Numerous alligator hunting lodges, guides, and outfitters cater to those willing to pay thousands of dollars to participate in an alligator hunt. Similarly, swamp tours that feature alligators as a main attraction bring in additional income. Unfortunately, a common practice by these tours is the illegal feeding of alligators to draw them close to the boat for a greater thrill for the tourists. Reputable tour guides do not feed wild alligators.

Chapter Eight

Captive Rearing of Alligators on Farms and Ranches

The very successful programs to protect and restore populations of the American alligator have been enhanced by the new industry of alligator farming and ranching, which has become extremely profitable since about 1980. The growth and popularity of these programs, as well as their economic value, have grown dramatically in recent years and now substantially exceed the value of wild alligator products.

Reports of "alligator farms" date back to the late 1800s,[1] though these early facilities were primarily tourist attractions in Florida, Arkansas, and California, with only minor effort devoted to alligator propagation. These early alligator "farms" generated income primarily through paid admissions, sales of merchandise, and sales and rentals of live alligators (mostly small alligators wild-caught and purchased from suppliers). One farm reported that less than one-twentieth of its income came from sale of hides taken from alligators raised in captivity. On farms where nesting did occur, hatching was successful only if eggs were left undisturbed in the nest. Attempts to incubate eggs resulted in total mortality,[2] possibly because staff did not handle the eggs properly. One farm in Jacksonville, Florida, was reported to contain twelve thousand alligators, which required about three tons of food daily, primarily fish or horse meat. Another in St. Augustine with six thousand alligators required between seven and eight hundred pounds of food daily. The latter, now named the St. Augustine Alligator Farm Zoological Park, has survived since its founding in 1893 by becoming a major Florida tourist attraction. Today, there are at least seven other alligator parks in Florida that cater to tourists. Another long-lasting alligator farm was founded in 1902 in Hot Springs, Arkansas (now known as the Arkansas Alligator Farm and Petting Zoo). The California Alligator Farm, established in 1907 in Los Angeles, was described in detail in an article by Arthur Inkersley in *Overland Monthly* in 1910 and again more than a

1. Kellogg, *The Habits and Economic Importance.*
2. Kellogg, *The Habits and Economic Importance.*

century later in the *Smithsonian Magazine*.[3] It moved to Buena Park in 1953 and closed in 1984.

Sadly, many of these tourist attractions invented new ways of abusing alligators, including allowing people to handle them, "ride" on them, and watch them slide down a sliding board. Many baby alligators were sold to tourists, who had little—if any—knowledge of how to care for them, resulting in many inevitable deaths or releases into inappropriate habitats. The concept of alligator "wrestling" also became a common attraction, a feature almost as fake as that of human professional wrestling. Few of these early "farms" served to increase populations of wild alligators. On a more positive note, they may have helped to educate people about alligator biology and their need for protection—although some of their biological statements were inaccurate and farfetched, such as the California farm's claim of a two-hundred-year-old alligator.[4]

Louisiana began a more modern and productive approach to raising alligators in captivity, for both commercial and conservation purposes, in the 1960s.[5] Prior to 1970, alligator farms were largely unprofitable because of high mortality rates of eggs and hatchlings, and most soon went out of business. However, extensive research by the LDWF in the 1970s and 1980s determined the proper methodology needed to successfully raise alligators in captivity, not only for commercial production, but also to help enhance wild populations.

Two distinct approaches to alligator farming have been utilized: true farming and ranching. Initially, farms were stocked by wild alligators from state-managed lands and retained in a closed "farming" system until this breeding stock became mature and produced eggs. Farmers then collected and incubated the eggs and raised the hatchlings until they reached a large enough size (4–5 feet [1.2-1.5 meters]) to provide commercially marketable skins. However, feeding captive alligators was expensive and the hatch rates of eggs produced by captive alligators were consistently low, making true farming impractical.[6] However, Ted Joanen and Larry McNease demonstrated that

3. Arthur Inkersley, "The California Alligator Ranch," *Overland Monthly* 56 (July–December 1910); Erin Blakemore, "When Kids Played with Alligators in Los Angeles," *Smithsonian Magazine*, November 5, 2015.
4. Inkersley, "The California Alligator Ranch."
5. Ted Joanen and Larry McNease, "Alligator Farming Research in Louisiana, USA," in *Wildlife Management: Crocodiles and Alligators*, eds. J. W. W. Grahame, et al. (Surrey Beatty & 1987); Ted Joanen and Larry McNease, "The Development of the American Alligator Industry," *Proceedings of the Intensive Tropical Animal Production Seminar, Townsville, Australia* (n.p., 1991).
6. Hines, "Alligator Harvest in Florida."

Alligator nest destroyed by raccoon at Manchac Wildlife Management Area, Louisiana,
August 20, 1987

time of egg collection (i.e., embryo age) and foods provided to mothers could affect the rate of successful hatching.[7] Hatching rates for eggs laid in captivity and artificially incubated initially ranged as high as 70 percent, although captive breeding success decreased over time.[8] Thus, the problems of limited production from true farms led to the development of alligator ranching.

Research on wild alligator populations conducted by biologists of the Louisiana Department of Wildlife and Fisheries and the US Fish and Wildlife Service demonstrated that 83 percent of eggs and hatchlings in the wild did not survive to reach 4 feet (1.2 meters) in length.[9] Predators such as raccoons destroyed many nests to eat the eggs. The small hatchlings were virtually defenseless and preyed upon by many other animals during the first few years after hatching. Because of this high natural mortality, the state's ranching program was initiated in 1986, permitting the collection of wild alligator eggs

7.   Joanen and McNease, "Alligator Farming Research in Louisiana, USA,"
8.   R. M. Elsey, et al., "Captive Breeding of Alligators and Other Crocodilians," *Crocodiles: Proceedings of the Second Regional Conference of the Crocodile Specialist Group, Darwin, Australia* (IUCN, 1994).
9.   Dave Taylor and Wendell Neal, "Management Implications of Size Class Frequency Distributions in Louisiana Alligator Populations," *Wildlife Society Bulletin* 12 (1984); Joanen and McNease, "The Development of the American Alligator Industry."

(and/or hatchlings) from nests on private lands. This provides a significant source of income for the landowner, as well as additional incentive to protect the wild alligator population and their habitat, since the landowner can sell the eggs and hatchlings to a rancher or raise and sell the resulting alligators on his own ranch. Ranching is now the preferred method for raising alligators in captivity (though still generally referred to as "alligator farming"). Maintaining the young alligators in controlled environmental chambers kept at a temperature of 85 degrees Fahrenheit (29.4 degrees Celsius) and providing nutritious foods increases growth rates so that a rancher can produce marketable-sized alligators within about two years (compared to three to four years in the wild).

Ranching also benefits the wild population because state law mandates that the rancher return to the area where the eggs were collected a percentage of the live alligators that have reached a size range of 3 to 5 feet (0.9 to 1.5 meters).[10] In recent years, Louisiana has required that 10 percent of the alligator hatchlings collected by the rancher be returned.[11] Survival rates for these rancher-hatched alligators is much greater than those hatchlings left to survive in the wild. During the period from 1986 to 2021, 1,288,166 alligators raised

Recently hatched American alligator at Manchac Wildlife Management Area, Louisiana, August 20, 1987

10. Joanen and McNease, "The Development of the American Alligator Industry."
11. *Louisiana's Alligator Management Program,* LDWF.

Controlled environmental chambers for young alligators at Rockefeller Wildlife Refuge, Louisiana, November 10, 1995

on Louisiana ranches (or an average over 35,000 per year) were returned to the wild in the areas where their eggs had been collected, replacing the estimated number that would have survived if the eggs had been left in the marsh.

Between 1972 and 1982, there were eight alligator farms (or "ranches") in Louisiana selling an average of about two hundred skins per year. The number of both farms and skins sold increased dramatically after 1982, reaching a total of 134 farms in 1991, selling a total of 118,976 skins.[12] The average length of skins sold was about 4.5 feet (1.4 meters). During subsequent years through 2019, the number of farms declined but eventually stabilized to about fifty or sixty large, efficient operations. Nevertheless, the number of skins sold continued to increase, reaching a high in 2018 of 450,221. Skin

12. Joanen and McNease, "The Development of the American Alligator Industry"; *Louisiana's Alligator Management Program*, LDWF.

value has varied dramatically during this period as the market price for skins has fluctuated, but has generally tended to increase, reaching the highest value in 2018 of $94,782,776. The value of alligator meat sold by the farms has also generally increased, reaching a high in 2018 of $9,454,641, making a total for the year of more than $104 million ($104,237,417), values much greater than those for the Louisiana wild alligator harvest that same year ($1,137,644 for skins and $3,247,531 for meat, or a total of $4,385,175). Thus, the alligator farming industry has become much more lucrative economically than the wild harvest—more than twenty-three times more valuable than the wild alligator harvest in 2018 alone. As seen in Table 10, during the fifty-year history of alligator harvesting in Louisiana (1972–2022), the number of wild alligators harvested was 1,135,197, with a value for hides and meat of $333,967,958. The comparable number of farm-produced skins (Table 11) was 8,326,635, with a value over one billion dollars ($1,398,937,608), more than four times the wild harvest. The Louisiana alligator ranching program has certainly been a significant asset to the state and has helped ensure the continued survival of the alligator population and its natural habitat.

Other states have implemented similar alligator ranching programs, in most cases following the lead and recommended methods of Louisiana. Still, Louisiana and Florida have dominated the industry with more than 98 percent of the production,[13] and the value of the Louisiana alligator industry, for both wild harvest and ranching, far exceeds that of all other states.

The development of the Florida program has been similar to that of Louisiana, beginning primarily in 1978.[14] Prior to that, there were only three alligator farms in Florida, and a fourth added between 1978 and 1980, with very few hides produced. In subsequent years, the number of farms increased rapidly, to stabilize at about fifty to sixty from 1990 to 2014.[15] The increasing availability of wild eggs and hatchlings for ranching stimulated another increase beginning in 2015, reaching a peak of ninety-one ranches in 2019, although only twenty-two of these were considered "active," meaning producing hides for sale. During the entire period (from 1977 to 2022) these ranches produced a total of 914,846, averaging about 19,000 per year. In the most recent period,

13. Mark G. Shirley and Ruth M. Elsey, "American Alligator Production: An Introduction," *Southern Regional Aquaculture Center Publication* no. 230 (September 2015).

14. Dennis N. David, "Summary of Alligator Farming Records in Florida," in *Proceedings of the Tenth Working Meeting of the Crocodile Specialist Group*, vol. 1 (IUCN, 1991).

15. David, "Summary of Alligator Farming"; Alligator Management Team, "Table 2. Estimated Producer Value of Alligator Harvests on Florida Farms During 1977–2022," Florida Fish and Wildlife Conservation Commission (2023), https://myfwc.com/media/1712/alligator-farm-value.pdf.

namely 2010 to 2022, the average was more than 29,000 per year. Over the entire period, the total hide and meat values reached more than $166 million and $42 million respectively, for a combined value of more than $209 million. The farm production value of alligators in Florida now exceeds the wild alligator harvest value there ($168 million) by more than $41 million.

Table 11. Ranch Alligator Harvest Data, Louisiana and Florida

Louisiana Farm/Ranch Harvest (1972–2022)				
Years	**Number of Hides Sold**	**Hide Value**	**Meat Value**	**Total Value 1972–2022**
1972–79	957	$71,977	$0	
1980–89	120,460	$15,272,216	$2,249,417	
1990–99	1,392,514	$91,875,893	$20,402,233	
2000–09	2,641,661	$332,927,188	$35,662,425	
2010–19	3,134,797	$624,131,272	$64,616,210	
2020–22	1,036,246	$189,967,611	$21,761,166	
Total	8,326,635	$1,254,246,157	$144,691,451	$1,398,937,608

Florida Farm/Ranch Harvest (1977–2022)				
Years	**Number of Hides Produced**	**Hide Value**	**Meat Value**	**Total Value 1977–2022**
1977–79	555	$35,561	$0	
1980–89	37,192	$6,177,709	$1,875,259	
1990–99	280,375	$28,932,407	$7,245,348	
2000–09	247,828	36,909,116	$7,765,217	
2010–19	246,786	$74,239,743	$18,115,584	
2020–22	102,110	$20,522,281	$7,757,007	
Total	914,846	$166,816,817	$42,758,415	$209,575,232

Sources: *Louisiana's Alligator Management Program 2021–2022 Annual Report*, Louisiana Department of Wildlife and Fisheries (2022); Alligator Management Team, "Table 2. Estimated Producer Value of Alligator Harvests on Florida Farms During 1977–2022," Florida Fish and Wildlife Conservation Commission (2023), https://myfwc.com/media/1712/alligator-farm-value.pdf.

Chapter Nine

Continuing Conservation Concerns for Crocodilians

T he various programs of the mid- to late twentieth century aimed at protecting American alligators and enhancing their populations have been a great success. Although pronounced endangered in 1967 by the US Fish and Wildlife Service, the American alligator had fully recovered by 1987 and was no longer considered endangered. In fact, in some locations, alligators are now abundant. As their populations have increased during the past fifty years, American alligators have once again provided a significant ecological benefit to wetland, bayou, and river environments, not to mention a substantial economic benefit for the southeastern states. In addition to the economic value of alligator hides and meat, alligator novelties such as alligator feet, heads, and tooth necklaces are now sold in many tourist gift shops. Substantial income can be realized by providing guided hunts for alligators. Swamp tours where alligators are the featured attraction have become popular. And a new industry of alligator farming has become extremely profitable since about 1980.

A healthy population of alligators at Payne's Prairie, Florida, February 18, 2010

However, some crocodilian species have not been so fortunate. Many of the other species are now threatened with extinction. Of the twenty-six crocodilian species currently recognized, seven are considered critically endangered, and four are listed as vulnerable, including the American crocodile (recognized as threatened by the US Fish and Wildlife Service). The remaining twelve are considered lower risk. Three other recently recognized species (West African crocodile, *Crocodylus suchus;* Congo dwarf crocodile, *Osteolaemus osborni*; and Central African slender-snouted crocodile, *Mecistops leptorhynchus*) have not been assessed.[1]

The American alligator's closest relative, the Chinese alligator, is one of those critically endangered. Its greatest problem is habitat loss, which is virtually complete. The remaining wild population of fewer than 150 is restricted to a small alligator reserve in Anhui Province of eastern China, consisting of several small ponds surrounded by agricultural fields.[2] Its survival will depend upon habitat restoration and reintroduction of captive-bred individuals.[3] There are several thousand in captivity worldwide, but captive breeding is not always successful. For long-term success, the small groups in the wild will require intensive management and protection, as well as the creation of additional habitat.

Today, the major factors reducing the populations of most crocodilians are habitat destruction and degradation and in some countries, the continuation of excessive killing (for hides, human consumption, sport, and out of fear). These and other factors causing endangerment of crocodilians, as well as many other endangered species, are almost always related to the uncontrolled growth of human populations. As human populations have increased, their competition for space with other species has risen in tandem, resulting in the destruction of many natural habitats. For most species, protection will not be sufficient for full recovery without proper restoration and maintenance of adequate critical habitat. Contributing to and exacerbating habitat loss is climate change, with effects such as wetlands and marsh deterioration, floods and droughts, sea level rise, saltwater intrusion, and increased frequency and strength of hurricanes. Reducing such effects will require major international efforts. Moreover, control of human populations is a critical, international problem, as is maintaining

1. Vliet, *Alligators.*
2. John Thorbjarnarson, et al., "Reproductive Ecology of the Chinese Alligator (*Alligator sinensis*) and Implications for Conservation," *Journal of Herpetology* 35, no. 4 (December 2001).
3. H. Jiang and X. Wu. "Alligator sinensis," IUCN Red List of Threatened Species 2018, https://dx.doi.org/10.2305/IUCN.UK.2018-1.RLTS.T867A3146005.en.

natural environments that will provide aesthetic and recreational benefits for humans.

Hopefully, with increased management, conservation and enhancement of habitats, and protective measures in the future, additional species can be upgraded to lower risk, just as occurred with the American alligator—an excellent example of what can be done to prevent the loss of endangered species. In most nations cooperation between government agencies and environmental organizations will be critical for their efforts to be effective.

We have all benefited from the recovery of the American alligator. The aesthetic attraction of this large, iconic reptile as part of our natural environment is something we can all appreciate. The opportunity to view large alligators in the wild is exciting. In the words of noted reptile biologist Dr. Archie Carr, "Having visible wild alligators in the landscape is a thing to be thankful for."[4] For the benefit of other species that need protection, we might modify Dr. Carr's quotation: "Having visible wild crocodilians in the landscape is a thing to be thankful for."

4. Archie Carr, "Eden Changes," in *A Naturalist in Florida: A Celebration of Eden*, ed. M. H. Carr (Yale University Press, 1994), 237.

References Cited

Books and Journal Articles

Alderton, David. *Crocodiles & Alligators of the World.* Facts on File, 1991.

Alexander, Michael. *Discovering the New World, based on the works of Theodor de Bry.* Harper and Row, 1976.

Allen, E. R., and W. T. Neill. "Increasing Abundance of the Alligator in the Eastern Portion of Its Range." *Herpetologica* 5, no. 6 (1949): 109–12.

Arthur, Stanley C. "The Alligator." In *Bulletin 18 (Revised): The Fur Animals of Louisiana.* Louisiana Department of Conservation, 1931.

Audubon, J. J. "Observations on the Natural History of the Alligator." *Edinburgh New Philosophical Journal* 2 (Oct. 1826–Apr. 1827): 270–80. Reprinted, *Louisiana Conservation Review* 2 (1931): 3–8.

Bartram, William. *Travels Through North & South Carolina, Georgia, East & West Florida, the Cherokee Country, the Extensive Territories of the Muscogulges, or Creek Confederacy, and the Country of the Chactaws; Containing an Account of the Soil and Natural Productions of those Regions, Together with Observations on the Manners of the Indians.* James & Johnson, Printer, 1791; repr., Dover Publications, 1928.

Blakemore, Erin. "When Kids Played with Alligators in Los Angeles." *Smithsonian Magazine* (November 5, 2015): 1–12.

Brandt, L., M. Campbell, and F. Mazzotti. "Spatial Distribution of Alligator Holes in the Central Everglades." *Southeastern Naturalist* 9, no. 3 (2010): 487–96.

Brunell, A., T. Rainwater, M. Sievering, and S. Platt. "A New Record for the Maximum Length of the American Alligator." *Southeastern Naturalist* 14, no. 3 (2015): N38–N43.

Bullen, Ripley P. "Some variations in settlement patterns in peninsular Florida." In *Southeastern Archaeological Conference, Bulletin No. 13*, edited by Bettye J. Broyles. Proceedings of the Twenty-Seventh Southeastern Archaeological Conference, 1971.

Bullen, Ripley P., and Adelaide K. Bullen. "The Palmer Site." *Florida Anthropological Society Publications,* no. 8 (1976): 1–55.

Butler, Ruth L. trans. and ed. *Journal of Paul du Ru*. Caxton Club, 1934.

Byington, Cyrus. *A Dictionary of the Choctaw Language*. US Government Printing Office, 1915.

Byrd, Kathleen M. "The Brackish Water Clam (*Rangia cuneata*): A Prehistoric 'Staff of Life' or a Minor Food Resource." *Louisiana Archaeology* 3 (1976): 23–31.

Byrd, Kathleen M. "Tchefuncte Subsistence: Information Obtained from the Excavation of the Morton Shell Mound, Iberia Parish, Louisiana." *Southeastern Archaeological Conference Bulletin* 19 (1976): 70–75.

Byrd, Kathleen M., and Robert W. Neuman. "Archaeological Data Relative to Prehistoric Subsistence in the Lower Mississippi River Alluvial Valley," In *Man and Environment in the Lower Mississippi Valley*, edited by Sam B. Hilliard. Louisiana State University, 1978.

Carr, Archie. "Alligators: Dragons in Distress." *National Geographic* 131, no. 1 (1967): 133–48.

Carr, Archie. "Eden Changes." In *Born of the Sun: The Official Florida Bicentennial Commemorative Book*, edited by. John E. Gill and Beth R. Rea. Florida Bicentennial Commemorative Journal, 1975.

Carr, Archie. Foreword to *The Alligator's Life History*, by Edward Avery McIlhenny. Society for the Study of Amphibians and Reptiles, 1976.

Chabreck, Robert H. "The American Alligator—Past, Present and Future." *Proceedings of the Annual Conference of Southeastern Association of Game and Fish Commissions* 21 (1967): 554–58.

Chabreck, Robert H., "Methods of Determining the Size and Composition of Alligator Populations in Louisiana." *Proceedings of the Annual Conference of Southeastern Associations of Game and Fish Commissions* 20 (1966): 105–12.

Challeux, Nicolas Le. *Discours de L'Historie de la Floride Contenant.la Cruaute des Espangnola, Contre les Sujects du Roy.* De Dieppe, 1566. https://archive.org/details/discoursdelhisto00lech/page/n7/mode/2up.

Charlevoix, Pierre Francois Xavier de. *Journal of a Voyage to North America*, vol. II. Translated by Louise P. Kellogg. Caxton Club, 1923.

Chesnel, Paul. *History of Cavelier de La Salle 1643-1687 Explorations in the Valleys of the Ohio, Illinois and Mississippi*, 1859. Translated by Andrée Chesnel Meany. G. P. Putnam's Sons, 1932.

David, Dennis N. "Summary of Alligator Farming Records in Florida." In *Crocodiles: Proceedings of the 10th Working Meeting of the Crocodile Specialist Group,* vol. 1. International Union for the Conservation of Nature, 1990.

Dennie, Bob. "Gatortime—Part Two." *Louisiana Conservationist* 25 (September-October 1973): 10–14.

Dowd, Elsbeth L. "Amphibian and Reptilian Imagery in Caddo Art." *Southeastern Archaeology* 30, no. 1 (2015): 79–95.

Dumont de Montigny, Jean-François-Benjamin. *The Memoir of Lieutenant Dumont, 1715–1747: A Sojourner in the French Atlantic.* Translated by Gordon M. Sayre. University of North Carolina Press, 2012.

Dutton, Harry J., Arnold M. Brunell, Dwayne A. Carbonneau, et al. "Florida's Alligator Management Program: An Update, 1987 to 2001." In *Crocodiles: Proceedings of the 16th Working Meeting of the Crocodile Specialist Group.* International Union for the Conservation of Nature, 2002.

Elsey, R. M., T. Joanen, and L. McNease. "Captive Breeding of Alligators and Other Crocodilians," In *Crocodiles: Proceedings of the 2nd Regional Conference of the Crocodile Specialist Group.* International Union for the Conservation of Nature, 1994.

Foster, William C., ed. *The La Salle Expedition on the Mississippi River: A Lost Manuscript of Nicolas de La Salle, 1682.* Translated by Johanna S. Warren. Texas State Historical Association, 2003.

Fundaburk, Emma Lila, M. D. F. Foreman, and V. J. Knight Jr. *Sun Circles and Human Hands: The Southeastern Indians Art and Industries.* University of Alabama Press, 2001.

Galloway, Patricia. "Sources for the La Salle expedition of 1682," In *La Salle and His Legacy,* edited by Patricia E. Galloway. University Press of Mississippi, 1982.

Gilliland, Marion S. *The Material Culture of Key Marco, Florida.* University Press of Florida, 1975.

Glasgow, Vaughn L. *A Social History of the American Alligator: The Earth Trembles with His Thunder.* St. Martin's Press, 1991.

Gravier, Jacques. "Journal of the Voyage of Father Gravier." In *Jesuit Relations and Allied Documents: Travels and Exploration of the Jesuit Missionaries in New France, 1610–1791,* vol. 66, Edited by Rueben G. Thwaites. Cleveland, OH: Burron Brothers Company, 1900.

Gremillion, Trent. "Alligator Effigy Mound." *Southwest Louisiana Archaeology* 1 (2019): 58–64.

Guevin, Bryan L. "The Ethno-Archaeology of the Houma Indians." PhD diss., Louisiana State University, 1983. https://digital commons.lsu.edu/gradschool_disstheses/8310.

Haag, W. G. "A Prehistory of the Lower Mississippi River Valley." In *Man and Environment in the Lower Mississippi Valley*, edited by Sam B. Hilliard. Louisiana State University, 1978.

Hines, Tommy, "Alligator Harvest in Florida," In *Amphibians and Reptiles: Status and Conservation in Florida*, edited by W. E. Meshaka Jr. and K. J. Babbitt. Krieger, 2005.

Hord, Ben McCulloch. "Alligator Shooting in Louisiana." *Trotwood's Monthly* 1–3 (1905): 250–252.

Hornaday, William T. *Our Vanishing Wildlife: Its Extermination and Preservation*. C. Scribner's Sons, 1913.

Howe, H. V., R. J. Russell, J. H. McGuirt, B. C. Craft, and M. B. Stephenson. "Reports on the Geology of Cameron and Vermilion Parishes." Bulletin 6, Dept. of Conservation, Louisiana Geological Survey, 1935.

Hutchinson, Dale L. *Bioarchaeology of the Florida Gulf Coast: Adaptation, Conflict, and Change*. University Press of Florida, 2004.

Inkersley, Arthur. "The California Alligator Ranch." *Overland Monthly* 56 (July–December 1910): 531–39.

Joanen, Ted. "Population Status and Distribution of Alligators in the Southeastern United States," In *Proceedings of the Southeastern Regional Endangered Species Workshop*, 1974.

Joanen, Ted, and Larry McNease. "Alligator Farming Research in Louisiana, USA," In *Wildlife Management: Crocodiles and Alligators*, edited by. J. W. W. Grahame, et al. Surrey Beatty & Sons, 1987.

Joanen, Ted, and Larry McNease. "The Development of the American Alligator Industry," In *Proceedings of the Intensive Tropical Animal Production Seminar*, 1991.

Joanen, Ted, and Larry McNease. "The Management of Alligators in Louisiana, USA," In *Wildlife Management: Crocodiles and Alligators*, edited by J. W. W. Grahame, et al. Surrey Beatty & Sons, 1987.

Joanen, Ted, Larry McNease, Guthrie Perry, David Richard, and Dave Taylor. "Louisiana's Alligator Management Program." *Proceedings of the Annual Conference of Southeastern Associations of Game and Fish Commissioners* 38 (1984): 201–11.

Joanen, Ted, Mark Merchant, Rebekah Griffith, Jeb Linscombe, and Angela Guidry. "Evaluation of Effects of Harvest on Alligator Populations in Louisiana." *Journal of Wildlife Management* 85, no. 4 (2021): 696–705.

Kellogg, Remington. *The Habits and Economic Importance of Alligators.* US Department of Agriculture Technical Bulletin 147. US Government Printing Office, 1929.

Langley, Ricky L. "Alligator Attacks on Humans in the United States." *Wilderness and Environmental Medicine* 16 (2005): 119–24.

Laudonnière, René Goulaine de. *History of the First Attempt of the French (The Huguenots) to Colonize the Newly Discovered Country of Florida."* In *Historical Collections of Louisiana and Florida, including Translations of Original Manuscripts Relating to Their Discovery and Settlement, with Numerous Historical and Biographical Notes.* Translated by Richard Hakluyt. Edited by B. F. French. New York: J. Sabin & Sons, 1869. Online facsimile edition at www.americanjourneys.org/aj-141/. Accessed November 14, 2024.

Le Page du Pratz, Antoine Simon. *The History of Louisiana, or of the Western Parts of Virginia and Carolina: Containing a Description of the Countries that Lie on Both Sides of the River Mississippi: With an Account of the Settlements, Inhabitants, Soil, Climate, and Products* (Paris: n.p., 1774; facsimile edition, Louisiana State University Press, 1975).

Liner, Earnest A. *The Culinary Herpetologist.* Bibliomania, 2005.

McIlhenny, E. A. *The Alligator's Life History.* Christopher Publishing House, 1935.

McIntire, William G. "Prehistoric Settlements of Coastal Louisiana." PhD Diss., Louisiana State University, 1954. https://digitalcommons.lsu.edu/gradschool_disstheses/8099.

McNease, Larry and Ted Joanen. "Distribution and Relative Abundance of the Alligator in Louisiana Coastal Marshes." *Proceedings of the Annual Conference of Southeastern Associations of Fish and Wildlife Agencies* 32 (1978): 182–86.

McWilliams, R. G., trans. and ed. *Iberville's Gulf Journals.* University of Alabama Press, 1981.

Milanich, Jerald T. "Fact or Fiction: Theodore De Bry's 1591 Engravings of Early Florida Indians." *Adventures in Florida Archaeology* (2016): 21–28.

Milanich, Jerald T. *Florida's Indians: From Ancient Times to the Present.* University Press of Florida, 1998.

Neuman, Robert W. and Nancy W. Hawkins. "Louisiana Prehistory." *Anthropological Study No. 6.* Louisiana Department of Culture, Recreation and Tourism, Louisiana Archaeological Survey and Antiquities Commission, 1987.

Nickum, Mary, Michael Masser, Robert Reigh, and John Nickum. "Alligator (*Alligator mississippiensis*) Aquaculture in the United States." *Fisheries Science & Aquaculture* 26, no. 1 (2018): 86–98.

North Carolina Alligator Management Plan, with July 12, 2018 Addendum. North Carolina Wildlife Resources Commission (2017). https://www.ncwildlife.org/media/1362/download?attachment.

Ouchley, Kelby. *American Alligator: Ancient Predator in the Modern World.* University Press of Florida, 2013.

Palmisano, A. W., Ted Joanen, and Larry I. McNease. "An Analysis of Louisiana's 1972 Experimental Alligator Harvest Program." *Proceedings of the Annual Conference of Southeastern Associations of Game and Fish Commissioners* 27 (1973): 184–206.

Pope, Clifford H. *Reptiles Round the World.* Knopf, 1957.

Quitmyer, Irvy R. "Zooarchaeological Remains from Bottle Creek." In *Bottle Creek: A Pensacola Culture Site in South Alabama*, edited by Ian W. Brown. University of Alabama Press, 2003.

Rees, M. A. *Archaeology of Louisiana.* Louisiana State University Press, 2010.

Rhodes, Walter E. "Conservation of a Dinosaur in Modern Times, South Carolina's Alligator Management Program," In *Proceedings of Eighth Eastern Wildlife Damage Management Conference,* 1997.

Rochford, Michael R., Kenneth. L. Krysko, Frank J Mazzotti, et al. "Molecular Analyses Confirming the Introduction of Nile Crocodiles, *Crocodylus niloticus* Laurenti 1768 (Crocodylidae), in Southern Florida, with an Assessment of Potential for Establishment, Spread, and Impacts." *Herpetological Conservation and Biology* 11, no. 1 (2016): 80–89.

Romans, Bernard A. *Concise Natural History of East and West Florida.* R. Aitkin, 1775; facsimile ed., University of Alabama Press, 1999.

Roosevelt, Theodore. *A Book-Lover's Holiday in the Open.* Charles Scribner's Sons, 1916.

Russo, Michael, and Irvy R. Quitmyer. "Developing Models of Settlement for the Florida Gulf Coast," In *Case Studies in Environmental Archaeology: Interdisciplinary Contributions to Archaeology*, edited by. E. J. Reitz, S. J. Scudder, and C. M. Scarry. Springer, 2008.

Shaffer, Lynda N. *Native Americans Before 1492: The Moundbuilding Centers of the Eastern Woodlands.* Routledge, 1992.

Shea, John G. *Discovery and Exploration of the Mississippi Valley with the Original Narratives of Marquette, Allouez, Membré, Hennepin, and Anastase Douay.* Joseph McDonough, 1903.

Shirley, Mark G., and Ruth M. Elsey. "American Alligator Production: An Introduction." *Southern Regional Aquaculture Center Publication*, no. 230 (September 2015): 1–4.

Sigler, L., J. Thorbjarnarson, F. O. Hinojosa, and B. Henley. "Searching for the Northern and Southern Distribution Limits of Two Crocodilian Species: *Alligator mississippiensis* and *Crocodylus moreletii* in South Texas, US, and in Northern Tamaulipas, Mexico." *Crocodile Specialist Group Newsletter* 26, no. 3 (2007): 6–7.

Smith, H. M. *Notes on the Alligator Industry.* Bulletin of the US Fish Commission for 1891. Washington, DC: US Government Printing Office, 1893.

Stevenson, C. H. *Utilization of the Skins of Aquatic Animals*, US Fish Commission Report for 1902. US Government Printing Office, 1904.

Stocker, Terry, S. Meltzoff, and S. Armsey, et al. "Crocodilians and Olmecs: Further Interpretations in Formative Period Iconography." *American Antiquity* 45, no. 4 (1980): 740–58.

Swanton, John R. *The Indians of the Southeastern United States.* Bureau of American Ethnology Bulletin 137. US Government Printing Office, 1946.

Swanton, John R. *Indian Tribes of the Lower Mississippi Valley and Adjacent Coast of the Gulf of Mexico.* Bureau of American Ethnology Bulletin 43. US Government Printing Office, 1911.

Swanton, John R. *Myths and Tales of the Southeastern Indians*. Bureau of American Ethnology Bulletin 88. US Government Printing Office, 1929.

Swanton, John R. *Source Material for the Social and Ceremonial Life of the Choctaw Indians*. Bureau of American Ethnology Bulletin 103. US Government Printing Office, 1931.

Taylor, Dave, and Wendell Neal. "Management Implications of Size Class Frequency Distributions in Louisiana Alligator Populations." *Wildlife Society Bulletin* 12 (1984): 312–19.

Taylor, Dave, Noel Kinler, and Greg Linscombe. "Female Alligator Reproduction and Associated Population Estimates." *Journal of Wildlife Management* 55, no. 4 (1991): 682–88.

Thorbjarnarson, John, Xiaoming Wang, and Lijun He. 2001. "Reproductive Ecology of the Chinese Alligator (*Alligator sinensis*) and Implications for Conservation." *Journal of Herpetology* 35, no. 4 (2001): 553–58.

Vliet, Kent. *Alligators: The Illustrated Guide to Their Biology, Behavior, and Conservation*. Johns Hopkins Press, 2020.

Weddle, Robert S., Mary Christine Morkovsky, and Patricia Galloway, eds. *La Salle, the Mississippi, and the Gulf: Three Primary Documents*. Texas A&M University Press, 1987.

Wicker, K. M., D. Davis, and D. Roberts. *Rockefeller State Wildlife Refuge and Game Preserve: Evaluation of Wetland Management Techniques*. Coastal Management Section, Louisiana Department of Natural Resources, 1983.

Williams, Lucy Fowler "The Calusa Indians: Maritime Peoples of Florida in the Age of Columbus," *Expedition* 33, no. 2 (1991): 55–60.

Woodward, Allan R., John H. White, and Stephen B. Linda. "Maximum Size of the Alligator (*Alligator mississippiensis*)." *Journal of Herpetology* 29, no. 4 (1995): 507–13.

Online Sources

Alligator Management Team. "Table 1. Estimated Producer Value of Wild Alligator Harvests in Florida During 1977-2022," Florida Fish and Wildlife Conservation Commission (2023), https://myfwc.com/media/1775/alligator-wild-value.pdf.

Alligator Management Team. "Table 2. Estimated Producer Value of Alligator Harvests on Florida Farms During 1977–2022," Florida Fish and Wildlife Conservation Commission (2023), https://myfwc.com/media/1712/alligator-farm-value.pdf.

Audubon Delta, "Paul J. Rainey Wildlife Sanctuary." Accessed November 21, 2024. https://la.audubon.org/conversation/paul-j-rainey-wildlife-sanctuary.

Barbee, Mark. *Alligator Hunt Orientation and Training Manual.* Arkansas Game and Fish Commission, 2020. https://www.agfc.com/en/hunting/big-game/alligator/.

Bry, Theodor de, *XXIV. Mode of Drying Fish, Wild Animals, and other Provisions.* 1591. State Archives of Florida, Florida Memory. https://www.floridamemory.com/items/show/294790.

Bry, Theodor de. *XXVI. Killing Crocodiles.* 1591. State Archives of Florida, Florida Memory. https://www.floridamemory.com/items/show/294792.

Elsey, R M., A. Woodward, and S. A. Balaguera-Reina. "*Alligator mississippiensis.*" IUCN Red List of Threatened Species, 2019. Accessed November 25, 2024. http://dx.doi.org/10.2305/IUCN.UK.2019-2.RLTS.T46583A3009637.en.

Fitzner, Zach, "In a Number of Different Cultures, Crocodiles Are Worshipped." Earth.com, March 28, 2019. https://www.earth.com/news/cultures-crocodiles-worshipped/.

Florida Fish and Wildlife Commission. "Alligator Bites on People in Florida." Updated December 2022. https://myfwc.com/media/1716/alligator-gatorbites.pdf.

International Union for Conservation of Nature's Red List of Threatened Species. https://www.iucnredlist.org/, particularly their Crocodile Specialist Group, http://www.iucncsg.org/pages/Conservation-Status.html. Accessed Fall 2024.

Jiang, H., and X. Wu. "*Alligator sinensis.*" IUCN Red List of Threatened Species, 2018. https://dx.doi.org/10.2305/IUCN.UK.2018-1.RLTS.T867A3146005.en.

"Louisiana's Alligator Management Program 2021–2022 Annual Report," Louisiana Department of Wildlife and Fisheries (2022). https://www.wlf.louisiana.gov/resources/category/annual-reports.

Seasholes, Brian, "The Great Gator Hoax." Green Watch, Capital Research Center, February 8, 2013. https://capitalresearch.org/app/uploads/2013/02/GW1302-final-1301181.pdf.

Texas Parks and Wildlife Department. "Alligators in Texas: Rules, Regulations and General Information, 2019–2020." Accessed December 2, 2024. https://tpwd.texas.gov/publications/pwdpubs/media/pwd_bk_w7000_1011.pdf.

Woodward, Allan R. "History of American Alligator Regulations in the U.S.A." Florida Fish and Wildlife Conservation Commission, June 12, 2007. https://myfwc.com/media/1742/alligator-regs-history.pdf.

Index

www.ingramcontent.com/pod-product-compliance
Lightning Source LLC
LaVergne TN
LVHW020616111025
823066LV00009B/100